T0332529

OCEAN RESOURCES

OCEAN RESOURCES

VOLUME II

SUBSEA WORK SYSTEMS AND TECHNOLOGIES

Derived from papers presented at the
First International Ocean Technology Congress
on EEZ Resources: Technology Assessment
held in Honolulu, Hawaii, 22–26 January 1989

edited by

DENNIS A. ARDUS

British Geological Survey, Edinburgh, U.K.

and

MICHAEL A. CHAMP

National Science Foundation, Washington DC, U.S.A.

KLUWER ACADEMIC PUBLISHERS
DORDRECHT / BOSTON / LONDON

Library of Congress Cataloging-in-Publication Data

International Ocean Technology Congress on EEZ Resources: Technology
  Assessment (1st : 1989 : Honolulu, Hawaii)
    Ocean resources / edited by Dennis A. Ardus and Michael A. Champ.
       p.   cm.
    "Derived from papers at the First International Ocean Technology
  Congress on EEZ Resources: Technology Assessment, Honolulu, Hawaii,
  1989."
    Includes index.
    Contents: v. 1. Assessment and utilisation -- v. 2. Subsea work
  systems and technologies.
    ISBN 0-7923-0954-5 (set : alk. paper). -- ISBN 0-7923-0952-9 (v. 1
  : alk. paper). -- ISBN 0-7923-0953-7 (v. 2 : alk. paper)
    1. Marine resources--Congresses. 2. Ocean engineering-
  -Congresses.   I. Ardus, D. A.   II. Champ, Michael A.   III. Title.
  GC1001.I58 1990
  333.91'64--dc20   90-5203

ISBN 0-7923-0952-9(I)
ISBN 0-7923-0953-7(II)
ISBN 0-7923-0954-5(set)

Published by Kluwer Academic Publishers,
P.O. Box 17, 3300 AA Dordrecht, The Netherlands.

Kluwer Academic Publishers incorporates
the publishing programmes of
D. Reidel, Martinus Nijhoff, Dr W. Junk and MTP Press.

Sold and distributed in the U.S.A. and Canada
by Kluwer Academic Publishers,
101 Philip Drive, Norwell, MA 02061, U.S.A.

In all other countries, sold and distributed
by Kluwer Academic Publishers Group,
P.O. Box 322, 3300 AH Dordrecht, The Netherlands.

Printed on acid-free paper

Printed in the Netherlands

DEDICATION

Dr. Kenji Okamura

On behalf of the participants and many friends who attended
the first International Ocean Technology Congress (IOTC), we would like
to honour posthumously Dr. Kenji Okamura, who was a Special Assistant
to the Minister for Science and Technology, a longtime Executive with
Mitsubishi Heavy Industries Ltd., and a Founding Director of the Japan
Marine Science and Technology Centre.

Dr. Kenji Okamura was internationally known and highly respected
for his distinguished career and many contributions to the advancement
of ocean science and technology for the development and utilization of
the oceans and their resources. Among his many accomplishments was his
distinguished service and valuable contributions as participant and
advisor to several Marine Technology Panels of the U.S.-Japan
Cooperative Program in Natural Resources (UJNR).

Dr. Okamura died on January 15th, 1989, one week before the IOTC
in Honolulu, Hawaii. He prepared several papers for this conference
which were presented by others and incorporated into the conference
record.

In his honour, we would like to dedicate this International Ocean
Technology Congress and the resultant conference papers to the memory
of Dr. Kenji Okamura.

Dr. Okamura will be remembered for his pursuit of the development
of the oceans for the benefit of mankind.

The International Ocean Technology Congress

PREFACE

Ocean engineering is generally considered to be concerned with studies on the effects of the ocean on the land and with the design, construction and operation of vehicles, structures and systems for use in the ocean or marine environment.

The practice of engineering differs from that of science in both motivations and objectives. Science seeks understanding of the principles of nature in terms of generalizations expressed as laws and classifications. Engineering seeks the application of knowledge of the physical and natural world to produce a benefit expressed as a device, system, material, and/or process.

From the standpoint of the financial sponsors of an engineering project, the ideal approach is one of minimal risk in which only proven knowledge, materials and procedures are employed. There is frequent departure from this ideal in anticipation of the increased benefit expected from a large increase in performance of a structure or device. The process of acquiring this new capability is engineering research.

Historically, ocean engineering developed with the application of engineering principles and processes to the design of ships and, later, to the machinery that propels them. In most societies, naval architecture and marine engineering are recognised as the origin of ocean engineering. In fact, the design of a ship constitutes the original systems engineering programme involving hydrodynamics/fluid flow, structural design, machinery design, electrical engineering and so on as well as requiring knowledge of the ocean environment (waves, corrosion, etc.).

The second historic period in the development of ocean engineering was the need to develop harbours and terminals to accommodate the increasing numbers of larger ships with heavier cargos. This involved designing and building structures to endure the ocean's attack and the construction of piers and supporting structures in unstable ground. This effort has expanded in magnitude and complexity as structures are built for use in deeper waters extending further offshore.

In recent years, offshore developments have created the need to develop highly sophisticated and reliable subsea work systems, both mobile and installed on the sea floor. Seabed completion systems for the offshore oil industry, ROVs and autonomous vehicles are prime examples.

Like all engineering efforts, ocean engineering is significantly influenced by advances in relevant sciences but the finding and sustainable theories developed by marine physicists, chemists, geologists and biologists are in turn often made possible by advances in technology. One may consider, for example, studies following the development of the towed magnetometer and the recognition from that data of the implications with regard to ocean spreading and the consequent development of the theory of global tectonics; or perhaps the knowledge of the processes giving rise to polymetallic sulphides in ocean ridge environments made possible by submersible expeditions; or the considerable advance in our understanding of the earth achieved through offshore and ocean drilling techniques.

Our awareness of resources, the ability to evaluate them and to appraise the environmental consequences of their exploitation depend on the interplay of science and engineering. The mutual dependance of the many disciplines involved in marine studies and projects is increasingly evident.

When we consider space, within our solar system only the earth has an atmosphere and ocean to support life. Therefore, the exploration, protection and development of ocean space and its resources must be given a major priority by mankind.

Michael A. Champ and Dennis A. Ardus

FOREWORD

The newly created Exclusive Economic Zone (EEZ) and its near equivalent in other countries help focus attention on the utility of the ocean. In a legal sense it formally extends a nation's economic interests out to 200 miles seaward. By so doing it brings the entire ocean closer to the thoughts and concerns of governments and their peoples.

How well we understand and constructively utilize such a wide range of adjacent oceanic variation may be a measure of how maritime oriented countries may be in tomorrow's world. We all recognise that a maritime nation has shallow water along its coasts. What is less recognised is that along some coasts and islands geologic activity has produced some very deep ocean trenches in EEZ areas. A very wide range of ocean depth, weather, ice, currents and biota fall within EEZ areas. These border areas are close to ports and population centres and hence are heavily traversed by the world's ships and aircraft.

This collection of papers delivered and discussed at the conference provides a current view of much of the technology and thinking about our capability of dealing with EEZ problems and opportunities. The variety of countries, organizations and participants illustrates the breadth of thinking around the world. The mix of ideas on capability and caution, of generality and detail and of electronics and mechanics shows that the participants have been realistically involved in working at sea and thinking about maritime problems. Subjects covered were manned, unmanned and autonomous platforms, along with essential supporting techniques and equipments for acoustic and optical search and classification. Robotics and their control systems were discussed as well as the equally important philosophy of measurement and methodology. Such a description of the state of the art in the late 1980's makes a good passport for entering the undersea technological world of the 1990's.

Rather than trying to summarize so many papers it may be more useful to encourage thinking about ways and directions that these technologies can best be applied to utilize the EEZ.

Occupation versus Transit

Because EEZ areas border coastlines and islands they are generally reasonably close to national ports. As such, the logistic and political problems relevant to extended operations may be simpler than for most extended high seas R & D. This suggests that special purpose craft and equipment can be designed to optimise more for capability on station than for high speed transit.

Data Transmission

It seems probable that a considerable number of EEZ stations will utilize fibre cables to the beach. This would permit high data rates and continuous TV that might be impractical or too expensive using antenna to provide a line of sight path seaward for several tens of miles.

Offshore Test Beds for Engineering Research
          Consider the EEZ as a handy area to economically develop, test and
debug new systems and train operators to improve reliability prior to
working in more distant waters where transit time is long or the
weather window is short.   Reliability is essential to develop
commercial funding and activity.   Also a country or a company may be
more willing to invest capital in permanent bottom facilities or large,
stable platforms moored offshore in their EEZ territory rather than out
in the high seas area.

Offshore Earthquake Research Facilities
          Lessening the chance of a major earthquake by initiating a series
of small ones is an obvious and intriguing possibility on theoretical
grounds but dubious at best near populated areas.   However, there may
be more isolated geologic features well offshore where initial
experiments bearing on this problem could be initiated.

          The sponsors and participants of this meeting clearly believed in
the value of quality of techniques and pluralism of choice.   I heartily
endorse their objectives and the printed results are significant steps
towards a better understanding and utilization of the ocean.

                              Allyn Vine
                              Woods Hole Oceanographic Institution,
                              Massachusetts, U.S.A.

ACKNOWLEDGEMENTS

The need for the International Ocean Technology Congress (IOTC) was recognised at a National Science Foundation and University of Hawaii sponsored conference in 1986 concerned with ´Engineering solutions for the Utilization of the Exclusive Economic Zone (EEZ) Resources`. This resulted in the establishment of a small international group of scientists and engineers with a common interest in the development and conservation of ocean space and resources which met at Heriot-Watt University, Edinburgh in 1987. Subsequently, the group has held planning meetings at the Energy and Mineral Research Organization in Taiwan and at the University of Hawaii in 1988.

The Congress, from which these papers were derived, was held in January 1989 in Honolulu, Hawaii. It was sponsored by:

National Science Foundation
Commission of European Communities
Institut Francais de Recherche pour l'Exploitation de la Mer
Industrial Technology Research Institute, Taiwan
University of Hawaii
Heriot-Watt University
Society for Underwater Technology
Marine Technology Society

Members of IOTC who have served as members of the IOTC editorial board for this volume include:

Dennis A. Ardus	British Geological Survey, U.K.
Norman Caplan	National Science Foundation, U.S.A.
Michael A. Champ	Environmental Systems Development Inc., U.S.A.
Chen-Tung A. Chen	National Sun Yat-Sen University, Taiwan
John P. Craven	Law of the Sea Institute, University of Hawaii, U.S.A.
Robin M. Dunbar	Heriot-Watt University, U.K.
Michel Gauthier	Institut Francais de Recherche pour l'Exploitation de la Mer, France
Jorgen Lexander	Swedish Defence Research Establishment, Sweden
C.Y. Li	Advisor on Science & Technology, The Executive Yuan, Taiwan
Kenji Okamura	Ministry of State, Science & Technology Agency, Japan
Boris Winterhalter	Geological Survey of Finland
Paul C. Yuen	University of Hawaii, U.S.A.

The considerable contribution of Fay Horie and Carrie Matsuzaki of the University of Hawaii in the organization of the first IOTC and in the preparation of this volume is gratefully acknowledged.
Pamela Pendreigh and Fiona Samson of Heriot-Watt University, Edinburgh are thanked for their preparation of the text for this Volume.

TABLE OF CONTENTS

PART I

Subsea Work Systems

ADVANCES IN MARINE ROBOTICS TECHNOLOGIES - STRATEGIC APPLICATIONS AND PROGRAMS

JAMES S. COLLINS
Department of Electrical Engineering
University of Victoria and
Royal Roads Military College F.M.O.
Victoria B.C., CANADA.

ABSTRACT. Recent Exclusive Economic Zone policies have given increased motivation for resource exploitation by many coastal nations. This activity will require a new generation of undersea equipment to support man in a hostile environment. This paper examines the importance of advanced marine robotics in subsea scientific, military and industrial applications. Development programs in Japan, the United States, Canada and Europe that are the most potentially supportive of the technology in each of these areas are discussed.

1. INTRODUCTION

As worldwide demand for greater quantities and types of natural resources increases, mankind grows more dependent on the potential of the oceans. To enable a more orderly future development of these resources, the United Nations Law of the Sea treaty was signed in December 1983 by 117 nations. The intent is that coastal countries have exclusive rights to the fish and other marine life up to 200 miles from their shore. Countries with a continental shelf as well have exclusive right to petroleum and other resources in the shelf up to 200 miles from shore (The Law of the Sea, 1986). Because of these conditions, some coastal countries increase their area of economic control greatly. For example in Japan's case, the increase is a factor of twelve (Ocean Development in Japan, 1986).

This increase in the area of sovereignty of ocean states gives them a proportionately greater incentive to procure new tools by which they can make these areas productive. Remotely operated vehicles (ROVs) are one such tool.

At this time, almost all commercially available submersible ROVs are free swimming vehicles which rely on a tether cable for power, communication and control. As the functionality of small size computers rapidly increases, it appears that vehicles with a low power requirement could be given a degree of autonomy. Such vehicles of this type, exhibiting a high degree of autonomy are advanced marine robots. The limits on autonomy are determined by the capability of navigation and telemetry systems in addition to the power and computer systems

3

D. A. Ardus and M. A. Champ (eds.), Ocean Resources, Vol. II, 3–10.
© 1990 *Kluwer Academic Publishers. Printed in the Netherlands.*

functionality. Aside from their environmentally oriented peripherals, much of their remaining technology will be generic to all advanced autonomous robots.

Countries such as Canada, France, Germany, Italy, Japan, Norway, Sweden, United Kingdom and the United States, are investigating the implementation of important advanced marine robot projects. So far one, if not the best funded project, is being conducted by Japan within the context of the Large Scale Project.

2. ADVANCED MARINE ROBOTICS - A STRATEGIC TECHNOLOGY

A technology can be called strategic if it has the potential for a long term effect on the ability to contribute or compete in applications of great scientific, industrial, military or economic importance. Advanced marine robotics rates high on all measures.

The scientific benefit of advanced marine robotics are found in the potential of such robots to extend at reasonable expense, mans' capability to position and operate scientific instrumentation in extremely hazardous or tedious undersea locations. Such instrumentation is generally involved with the measurement of chemical, physical and biological oceanographic parameters as a function of time and/or position. This statement includes undersea exploration and survey activities which are particularly important in characterizing natural resources occurring within EEZs.

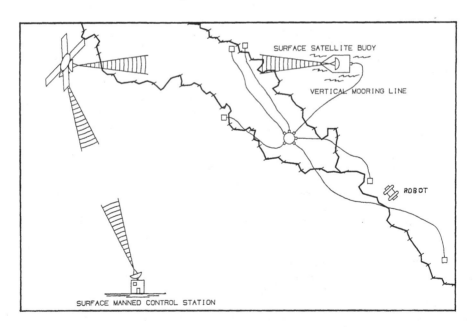

Figure 1. Robot based remotely managed scientific system for hot-vent study.

Of current scientific interest for example, is the study of activity near the "hot vents" occurring in ocean basins. Figure 1 shows a potential system for remote management of unattended advanced marine robots and for basic measurement and maintenance activity in such an area. Direct control of the robots can be implemented through a coupled system using short range acoustics, a sea bottom network, a deep sea buoy and satellite communications. Ideally the robot would remain at the work site unattended. The most difficult aspects to this system would be to provide:

-   power supply replenishment at the main central node.

-   real time communication to the vehicle by connecting it to the network by direct coupling or short acoustic link.

-   vehicle guidance between work site by an automated long based system.

Military sovereignty and surveillance can be potentially implemented by a similar remotely managed system (see Fig. 2). Other military applications will be found in mine countermeasures, salvage, rescue and hull maintenance.

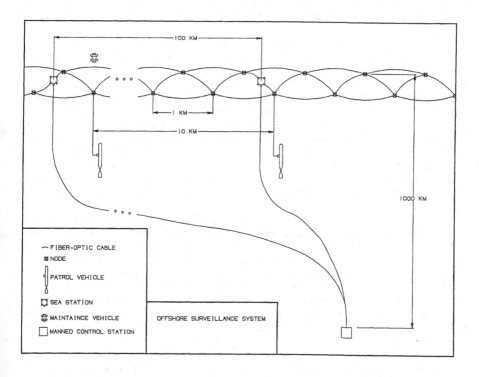

Figure 2.   Robot based remotely managed subsea surveillance system.

The main existing industrial application of marine robotics using tethered vehicles is in production of offshore petroleum. It is anticipated that much of this activity will be assumed by advanced untethered vehicles. This will happen when their operational characteristics and reliability surpass those of tethered vehicles. Industrially oriented operations include:

- exploration
- manipulation of simple valves, levers, etc.
- search and recovery
- cleaning, welding and non-destructive testing
- inspection and monitoring
- assembly

Also as demand rises for manganese and related minerals, the use of advanced marine robotics in the support of underwater mining will increase.

It is as well important to note the importance of marine robotics as a potentially prolific source of spinoff technology. We have seen the invention of something conceptually as simple as the telephone catalyze developments in other areas such as microwaves, computers and transistors. Similarly, marine robotics will profit from, and motivate progress in many technologies including those listed in Table 1. The spinoff effect could be considerable for non-marine robotics and other devices.

Advanced marine robots will see useful technological progress sooner than all mobile land robots except the very simplest. Some subsystems in marine robots differ greatly from their equivalent subsystems in land mobile robots. These subsystems relate to motion control (navigation, path planning and locomotion) and communications. In untethered robots, marine robots have some difficulties with navigation and communication. Land robots have much greater problems with path planning and locomotion. Thus, marine vehicles are likely to be more useably advanced and more economically rewarding sooner than land vehicles. Also, for countries with long coastlines and great marine resources, development of marine robots is more relevant than land mobile robots. The latter are of more equal interest to all people.

Strategic projects with easy-to-understand goals are valuable in gaining public support for new technology development. If people can understand easily the benefits of a marine robot then funding, graduate student interest, etc., are more readily available for the component technologies when presented as a finished product. Marketing of the individual esoteric component technologies would be likely less rewarding. If marine robotics were marketed publicly with the same enthusiasm and success as space robotics, politicians would be able to justify marine strategic projects with a similar robotic research and development payout but at a much reduced cost.

From an economics point of view, it seems clear that the development of advanced marine robotics for scientific, industrial and military applications represents a multifaceted opportunity.

TABLE 1. Technologies for advanced marine robotics implementation

Computer Hardware
- architectures for multiperipheral
  robots
- digital signal processing
- application specific integrated
  circuits

Computer Software
- robotic software and languages
- AI for mobile robot guidance
- computer aided design
- algorithm and program design
- simulation

Telemetry
- acoustic (channel is multipath,
  doppler and bandwidth limited
  with a time delay)
- fibre optic
- laser

- motion compensated video
  compression
- bandwidth efficient communication

Propulsion
- thrusters (low noise,
  high efficiency,
  omni-directional)

Manipulators
- teleoperation
- salt water actuators
- fingers, hands, wrists, arms
- impedance control

Imaging
- acoustic, optical
- pattern recognition

Sensors
- optical
- tactile
- acoustic
- other

Energy Sources
- high density closed
  cycle
Reliability

Control and Hydrodynamics
- vehicle response and stability
  in transit and on work site in
  high and fluctuating current
  situations
- vision/manipulator system
  co-ordination
- multivehicle systems

Human Factors
- man/machine interface

Structural
- pressure housing materials and
  design

Navigation
- inertial, magnetic, pressure
  sensors
- fibre optic gyros
- acoustic(short/long baseline,
  doppler
- object recognition
- obstacle avoidance

Modelling
- robot dynamics
- communication

3.   AREAS OF LARGE SCALE ACTIVITY IN ADVANCED MARINE ROBOTICS

As stated previously, activity in advanced marine robotics is confined
to several countries. If one considers only large scale non-military
research and development, there appear to be only four areas which
presently show the potential for significant contributions to advanced
marine robotics in the next several years.

3.1. The Japanese Large Scale Project and Undersea Robotics

Prior to the mid 1980s, marine robotics did not receive a great amount
of attention in Japan. However, the area will likely be another
example of a technology to be accelerated by the Japanese Ministry of
International Trade and Industry's (MITI) National Research and
Development Program (popularly known as the Large Scale Project). No
doubt the most expensive marine robot ever constructed (approx. $50
Million US), a MITI vehicle intends to break new ground in autonomous
and remote controlled operation, acoustic imaging, underwater
telemetry, omni-directional thrusters, sea water actuators, ceramic
applications, underwater navigation, positioning and control. This
project's immediate goal is to assist in the performance of dangerous
undersea work normally performed by divers in support of offshore oil
exploration. This work involves inspection, maintenance and operation
around drilling rigs usually located in 200 m or less water depth. The
operating conditions involve water temperatures between $-2^{\circ}C$ and $30^{\circ}C$,
currents less than 2 knots, waves up to 3 m and water transparency of
at least 5 m.
        The national project structure achieves what it sets out to do for
several reasons. Projects selected are relevant to the country. Goals
have a technically reasonable chance of being reached. The MITI
project organization structure amounts to a well coordinated and
motivated team effort among the many participants which include both
universities and industry. Sufficient financial backing is allocated
to achieve the goals which private industry could not independently
afford because of high cost, long term absence of profit and high risk.
        The selection of appropriate national projects seems to be a
catalyst for the achievement of very impressive goals. There are many
other facets to the conduct of high technology business in Japan
(Abeggten and Stalk, 1985) but without the contribution of the Large
Scale National Projects, Japan would probably be much more constrained
to marketing current technology as opposed to being the leader it is.
There is little doubt that some substantial technological gains will be
made based on this project which is due to finish in 1990.

3.2. The EUREKA EU-191 Project

This project is similar in motivation to the Japanese MITI project.
The funding is similar ($60 M US) but the period is shorter (5 years)
and two vehicles will be built. One vehicle will be a tethered semi-
autonomous work and inspection robot. The other will be an untethered
robot for underwater survey. Seventeen organizations from Italy,
England and Denmark are participating. Funding was approved in late
1987, so it is too early for results. Proposed goals are described
(Bevilacqua, et al, 1988).

3.3. Other Possibilities

So far there is no strategic research program explicitly for advanced
marine robotics in either the United States or Canada. One possibility
in the United States is the Engineering Research Center (ERC) program

of the National Science Foundation. However, as Table 1 illustrates, the breadth of technology required would involve the participation of more than one centre. A network of centres would be more appropriate if it could be developed in the context of the ERC program discussed (Suh, 1987).

In Canada, a recent federal government initiative has established a Networks of Centres of Excellence program. The first competition for funds closed in November 1988. One application was from a consortium of Canadian universities interested in advanced robotics research. If successful, potential funding could be in the order of $20 M US. There is not particular interest in any application area at this time although marine, space and terrestrial mobile robots are obvious possibilities.

Also in Canada, a consortium of companies in British Columbia has proposed to build an untethered free-swimming vehicle with manipulators. A contractor as been selected to further develop the proposal.

4.   CONCLUSION

This paper has outlined the long term importance of advanced marine robotics as a strategic technology. A summary of important activity in Japan, Europe and North America is given.

5.   REFERENCES

Abeggten, J.C. and Stalk, G.C., Jr. (1985) "Kaisha, the Japanese Corporation", Basic Books Inc., New York.

"Advanced Robot Technology" (1988) Advanced Robot Technology Research Association, Tokyo.

Bevilacqua, S. et al (1988) "Eureka EU-191 Project: Advanced Underwater Robots", Second International Workshop on Subsea Robotics, Joint Coordinating Forum for the International Advanced Robotics Program, Tokyo.

Collins, J.S. (1987) "Japanese Marine Robotic Advances and Implications for Canada", Marine Engineering Digest, Vol. 6, No. 4, pp.4-20.

"National Research and Development Program" (1985) Japan Industrial Technology Association, Tokyo.

"Ocean Development in Japan" (1986) Science and Technology in Japan, pp.15-19, October-December.

"Presentation of Results of Limited Operations Robot Research and
Development" (1987) Advanced Robot Technology Research
Association, Tokyo (translation available from NITS Order No.
PB88-140033).

Suh, N.P. (1987) "The ERCs: What We Have Learned", Engineering
Education, pp.16-18.

"The Law of the Sea" (1986) Encyclopedia Britannica, Micropaedia,
Vol. 10, Pg.577.

DESIGN CONSIDERATIONS FOR UNDERWATER ROBOTIC SYSTEMS

TYLER SCHILLING
Schilling Development, Inc.
Davis
California
U.S.A.

ABSTRACT. In the past several decades, researchers have learned a
great deal about the diverse field of robotics as well as the design of
mechanisms for underwater use. The sea imposes an aggressive
environment on machines as complex as the typical robot. This less
than ideal environment requires a special blend of forward ideas
tempered with a practical design approach. Due to the exciting
conceptual nature of these machines, the fundamental requirements are
sometimes given inadequate consideration.

A large percentage of subsea robotic equipment experiences failure
in initial application. Inadequate attention to the more mundane
details of design - seals, bearings and structure for example - rather
than poor conceptual design causes the majority of these malfunctions.
These types of failures provide a very frustrating distraction from the
research at hand.

A successful machine design begins with the segregation of the
design elements into two categories. The first category - truly new
and unique equipment - is the sole subject of the research. The second
category is the group of design concepts utilized previously and not
requiring additional development. The designer is advised to place as
much emphasis on the second category as possible.

1.   THE SUBSEA ENVIRONMENT

To improve the probability of success, the designer must attempt to
understand the subsea environment as well as possible. Some of the
more important aspects of working in an ocean medium are:

- Salt water corrosion
- High hydrostatic pressure
- Physically abusive conditions
- Limited maintenance facilities
- Increased mission criticality

These elements must be weighted differently for production or
research applications. The primary reason for this is that a

11

D. A. Ardus and M. A. Champ (eds.), Ocean Resources, Vol. II, 11–17.
© 1990 Kluwer Academic Publishers. Printed in the Netherlands.

production application is expected to have a longer useful life than a
developmental tool.

The first aspect of the sea requiring attention in design is the
problem of salt water corrosion. This should first be addressed by
careful selection of construction materials. This design challenge can
be nearly eliminated by selection of materials which are inherently
stable in seawater. Figure 1 shows relative corrosion resistance of
some common structural materials. In addition to reliable base
materials, coatings can be used to enhance surface corrosion
properties. A common example of this is the use of hard anodizing on
aluminium components. This technique works well as long as the coating
remains intact.

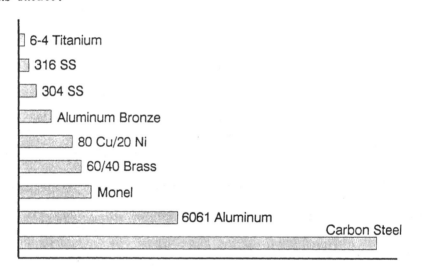

Figure 1.   Relative Seawater Corrosion Rates of Common Metals.

Another important corrosion consideration is the selection of
compatible materials to prevent the occurrence of undesirable galvanic
couples. Figure 2 shows the relationship of different metals in the
galvanic series. The designer must minimise the surface area of highly
noble metals coupled to less noble metals.

Plastics may be used to avoid the corrosion problem, however, the
designer must fully understand the water absorption characteristics of
candidate plastics. One exception in the area of non-metallics usage
is carbon reinforced plastics. Carbon fibre is at the extreme end of
the galvanic series and can cause accelerated corrosion of less noble
materials.

Component form factors also play an important role in corrosion
resistance. Seal placement, thread design and areas that promote water
capture and entrapment are illustrated in Figure 3. A subtle change in
the relationship of these design elements can change a well behaved
design into a performance nightmare.

Material	Electrode Potential, Volts	Material	Electrode Potential, Volts
Graphite	+0.25	Copper Alloy 706 (90Cu-10Ni)	-0.28
Platinum	+0.15	Copper Alloy 442 (Admiralty Brass)	-0.29
Zirconium	-0.04	G Bronze	-0.31
316 Stainless Steel (Passive)	-0.05	Copper Alloy 687 (Aluminum Brass)	-0.32
304 Stainless Steel (Passive)	-0.08	Copper	-0.36
Monel 400	-0.08	Alloy 464 (Naval Rolled Brass)	-0.40
Hastelloy C	-0.08	410 Stainless Steel (Active)	-0.52
Titanium	-0.10	304 Stainless Steel (Active)	-0.53
Silver	-0.13	430 Stainless Steel (Active)	-0.57
410 Stainless Steel (Passive)	-0.15	Carbon Steel	-0.61
316 Stainless Steel (Active)	-0.18	Cast Iron	-0.61
Nickel	-0.20	Aluminum 3003-H	-0.79
430 Stainless Steel (Passive)	-0.22	Zinc	-1.03
Copper Alloy 715 (70Cu-30Ni)	-0.25		

Figure 2.   Galvanic Series in Flowing Water.

The second environmental consideration in the design process is the effects of hydrostatic pressure. For rough design calculations, hydrostatic pressure in pounds per square inch at a given depth approximately equals one half the depth in feet.

The designer must carefully consider all cavities within the subsea robot that must resist hydrostatic pressure. These are typically housings which contain control system electronics or instrumentation. It is good design practice to minimise the volume of these housings, which in turn will reduce the surface area, weight and complexity. Housing stability is a feature that must be physically tested to ensure that no failures occur.

The application of seals to resist hydrostatic pressure also requires close examination in the design process, as shown in Figure 3. A face seal allows surface pressure to close the seal extrusion gap while a diametrical seal maintains a relatively constant gap. Additionally, a face seal is easier than a diameter seal to assemble and disassemble. These types of seals are considered static seals - the surfaces that the seal acts on are stationary relative to one another. O-rings are an excellent choice for these static seals, but should only be considered for dynamic seal applications in special cases. For dynamic seal applications, in which the elements are in motion relative to each other, there are a number of seal designs that are suitable. Most of these consist of a hard synthetic material with good wear properties which is energised with a spring. Figure 4 shows examples of good dynamic seals.

Internal fluid dynamics is an important design consideration in systems which contain fluids. In the factory environment at one atmosphere, fluids behave in a specific, predictable manner, and the

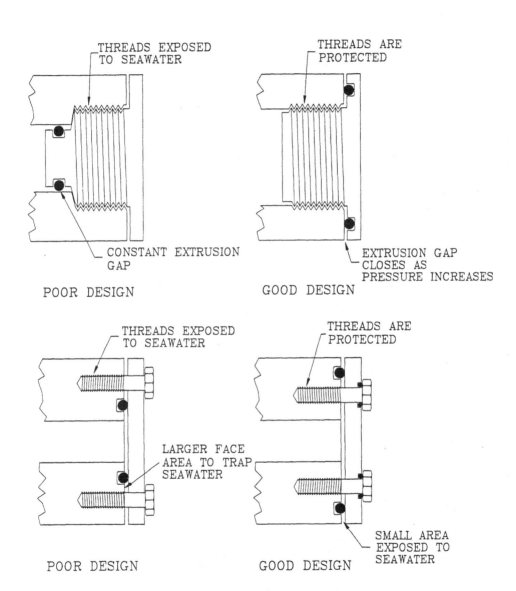

Figure 3.   Seal and Thread Design Applications.

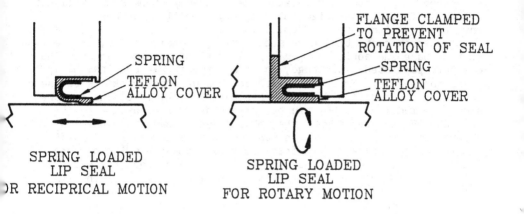

SPRING

TEFLON
ALLOY COVER

SPRING LOADED
LIP SEAL
OR RECIPRICAL MOTION

FLANGE CLAMPED
TO PREVENT
ROTATION OF SEAL

SPRING

TEFLON
ALLOY COVER

SPRING LOADED
LIP SEAL
FOR ROTARY MOTION

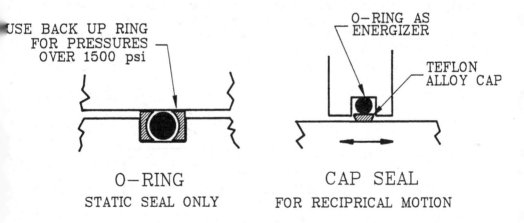

USE BACK UP RING
FOR PRESSURES
OVER 1500 psi

O—RING
STATIC SEAL ONLY

O—RING AS
ENERGIZER

TEFLON
ALLOY CAP

CAP SEAL
FOR RECIPRICAL MOTION

Figure 4.   Dynamic Seal Design Applications.

majority of fluidic system experience has been built on this one atmosphere behaviour. Some designs are only possible because the vapor pressure of the fluid can easily be reached and transient volume changes are easily accommodated. These same designs can fail catastrophically at depth because the high ambient pressure will not allow the fluid to vaporize, causing high positive and negative pressure transients to develop.

To accommodate the third subsea design consideration, the designer must prepare the equipment for a great deal of physical abuse. The ocean is populated with large, heavy equipment that requires heavy hands to operate. Equipment that requires watchmaker skills to use or maintain should be avoided. If this cannot be avoided, the components should be modular so they can be interchanged with ease.

Launch and recovery of the robot presents dangerous and potentially damaging conditions that require some consideration in design. The equipment will inevitably be suspended swinging from the side of a ship, crashing into the water with every ocean swell. Electronics and control cavities must be opened regularly in high humidity salt air environment.

Limited maintenance facilities is a fourth consideration in the design of subsea robotic equipment. Limited laboratory and repair facilities shipboard or on oil platforms should lead the designer to a minimum of special tools and calibration equipment required to maintain the system. The designer should attempt to use large fasteners wherever possible. Here also, modular designs provide relief from detailed repair and maintenance procedures.

The final element to be considered in subsea design is increased importance of mission success. Typically, the robotic equipment is at the focus of a very expensive operation involving large crews and costly support equipment. Figure 5 illustrates that it is less costly to thoroughly think through and verify a design before it is put to use. Once the equipment is at sea, the cost of an overlooked detail of design is extremely high.

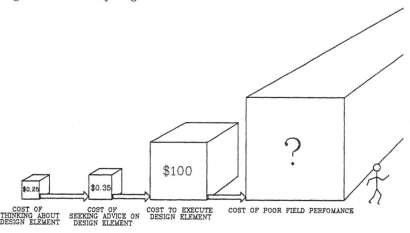

Figure 5.   Relative Costs in the Design Process.

2. SUMMARY

The designer should carefully segregate the robotic design elements
into developmental items and previously utilized items. Generally, it
is the more basic design details that trip up an otherwise viable
concept. For this reason, the designer should not close out the less
interesting portions of the design too early in the design process.
The designer of subsea robotic equipment should never forget this
formula:

GOOD CONCEPT + POOR TECHNIQUE = A BAD EXPERIENCE

The author does not intend to discourage anyone from designing
robots for the subsea environment, but rather to emphasize the need for
attention to the small details of design which are so often overlooked.

3. BIBLIOGRAPHY

Donachie, Matthew J., Jr. (ed) (1982) "Titanium and Titanium
    Alloys", Metals Park, Ohio: American Society of Metals.

Parker Seal Group, O-Ring Division (1982) "Parker O-Ring
    Handbook", Lexington KY.

Parmley Robert O. (1977) "Standard Handbook of Fastening and
    Joining", New York, NY, McGraw-Hill Book Co.

Titanium Metal Corporation of America, TIMET Division "Corrosion
    Resistance of Titanium", Pittsburgh, PA.

CONTROL CAPABILITIES OF JASON AND ITS MANIPULATOR

DANA R. YOERGER and DAVID M. DiPIETRO
Deep Submergence Laboratory
Department of Ocean Engineering
Woods Hole Oceanographic Institution
Woods Hole, MA 02543,
U.S.A.

ABSTRACT. Underwater work systems can be made more effective if
vehicles and manipulators are automatically controlled and can work
together in a coordinated fashion. Coordinated control can greatly
increase the workspace of the manipulator without the need for
additional or larger manipulator links. Such control will also
minimize the need for attachment points, which is especially important
in unprepared or poorly structured environments.
   This paper reviews two major elements of the JASON system that
will support such control. The first element is a hover control system
for JASON, and results from a recent experiment in shallow water are
presented. The second element concerns the design of a manipulator
that is especially suited to tasks requiring force control that will
fit very naturally into a combined vehicle-manipulator control system.
The overall philosophy and design of the arm are summarized.

1. INTRODUCTION

ARGO/JASON is an underwater vehicle system that will support scientific
operations to 6000 m. ARGO is a towed system that performs
simultaneous acoustic and optical imaging of the seafloor. Designed
primarily for geological survey, ARGO was used to locate the TITANIC in
1985. In the near future, ARGO will also serve as a garage for JASON.
After an area of interest has been located with ARGO, JASON will be
deployed to perform close-up inspection and manipulative work. JASON
Jr., a smaller forerunner of JASON, was operated from the submersible
ALVIN and was used to explore inside the TITANIC.
   ARGO/JASON has an infrastructure of computers, telemetry, and
power systems that can support sophisticated scientific sensors as well
as advanced control of JASON and its manipulators. At the core of this
infrastructure is a steel-armoured cable that provides power and fibre
optic telemetry to ARGO to depths of 6000 m. The cable contains copper
conductors that deliver approximately 10 KVA as well as three single
mode optical fibres (Von Alt, 1988). The telemetry system provides
multiple high-quality video channels from both ARGO and JASON to the
surface as well as 10 full duplex telemetry channels each with a

D. A. Ardus and M. A. Champ (eds.), Ocean Resources, Vol. II, 19–29.
© 1990 *Kluwer Academic Publishers. Printed in the Netherlands.*

capacity of 10 megabits/s. A local area network, optimized for real-time response, will provide high bandwidth communication between computers on the surface, in ARGO and in JASON.

A goal of the JASON effort is to provide precise, automatic control of the vehicle as well as coordinated control of the vehicle and its manipulator all under the supervision of the vehicle pilot. Such a capability can increase the range of tasks that can be performed and decrease completion time. This is particularly important for many scientific tasks, where the environment is usually not prepared. In scientific sampling, for example, the ability to stop and take a core sample without landing or attaching can speed up the sampling process and reduce the disturbance of the bottom.

This paper reviews two major elements of the JASON system that will support the overall control scheme: a hover control system and a high performance manipulator. These elements are necessary but not sufficient to achieve good coordinated control of the manipulator and vehicle. Another important topic under consideration concerns the design of control algorithms that can deal with the dynamics of the vehicle and manipulator as a unified system, and this issue will not be addressed here.

The first element presents results from a recent experiment with a hover control system for JASON. Experimental performance of the precision navigation system is summarized, a dynamic model identification is presented, and performance of the closed-loop system is described.

The second element presented in this paper concerns the design of a manipulator that is especially suited to tasks requiring force control. This arm will fit naturally into a combined vehicle-manipulator control system and into a future dual manipulator configuration. The overall philosophy and design of the arm are summarized.

2.   CONTROL OVERVIEW OF JASON

The JASON vehicle is shown in Figure 1. The vehicle is powered electrically with each of its 7 thrusters driven by a 1/3 hp brushless dc motor. The motors, their controllers, and much of JASON's power system run at ambient pressure in oil-filled volumes to save space and weight.

JASON was designed to be a controllable platform. The separation of the centers of buoyancy and gravity was made as large possible to provide good passive stability in pitch and roll. The thrusters were placed to provide uncoupled forces and moments in translation and heading while minimizing their effect on pitch and roll.

JASON is equipped with sufficient sensors to permit good control. Heading and heading rate are measured by a flux-gate compass (ENDECO) and a directional gyro (Humphries). Pitch and roll are measured with pendulum-type sensors (Schavitz). Depth is measured by an absolute pressure transducer (Paroscientific) while altitude is measured by a 200 kHz acoustic altimeter (Datasonics).

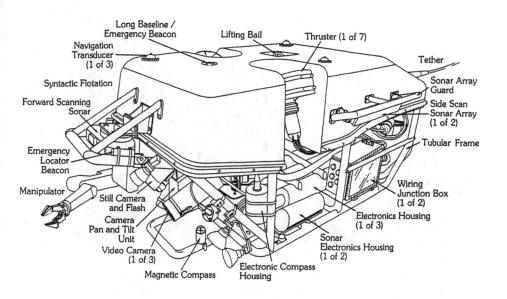

Figure 1. JASON is a remotely operated vehicle designed to perform
scientific tasks to depths of 6000 m. It has been designed to support
advanced manipulation and control concepts and a variety of scientific
instruments.

The vehicle is equipped with acoustic navigation transceivers
which are controlled and interrogated through the telemetry system.
The navigation system, called SHARPS (SHARPS User Guide), uses
extremely broadband pulses centred at 300 kHz to provide positioning on
the order of centimetres over a maximum range of approximately 100 m.
In dockside tests, two of these transceivers mounted on the
vehicle were used in conjunction with a three element net of
transceivers hardwired back to the control computers as shown in
figure 2. The two transceivers on the vehicle permit position and
absolute heading to be determined. Unlike the measurement obtained
from the flux-gate compass, the heading estimate from the navigation
system was automatically registered with the coordinates of the net and
was free from the large magnetic anomalies found around the dock.
The short-term repeatability of SHARPS range measurements is
summarized in figure 3. The measurements were made by logging ranges
from the vehicle to the net transceivers with JASON parked on the
bottom or by ranging between fixed elements of the net. Each point
represents the standard deviation of about 200 samples taken at a rate
of 5 per second. Short-term repeatability exceeds 1 mm (1 σ) and is
surprisingly independent of range. These tests were limited in range
due to the short length of JASON's tether that was used in these
preliminary tests, and the repeatability undoubtedly degrades as the
maximum range is approached. Additionally, in obtaining a cartesian
solution, these range errors are amplified by net geometry.

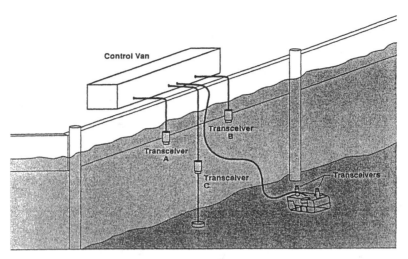

Figure 2. Test setup for JASON off the Woods-Hold dock. A hardwired
high frequency acoustic navigation system is used to navigate JASON
precisely and to obtain periodic absolute heading references.

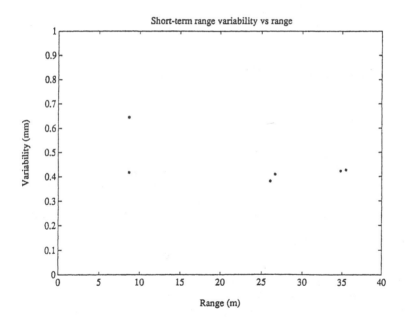

Figure 3. This plot shows the short-term variability of the acoustic
range measurements between fixed transceivers. The variability is very
low and substantially independent of range, although degradation of
precision is expected as the maximum range of 100 m is approached.

These results indicate that sufficient position measurements can be obtained to support combined good vehicle control and even vehicle-manipulator control. The current SHARPS implementation requires hardwired net transceivers, which permits operation in high multipath environments such as tanks and around structures in shallow water applications. A transponding equivalent would permit practical application in deep water.

3. DYNAMICS AND CONTROL FOR HOVER

A hover control system was designed and implemented for JASON and tested off the Woods Hold dock in the configuration shown in figure 2. The setup was used first to perform a system identification and then to test the control system.

System identification was performed for three translations and heading. Each axis was driven individually by a force or moment command consisting or filtered white noise. The position was obtained from SHARPS at approximately 5 samples per second. These positions were transformed incrementally into body referenced displacements, which were then differentiated and smoothed to obtain the body referenced velocities.

Figure 4 shows an example identification run for forward motion. The top plot shows the commanded input force. In this case, the maximum commanded force is about half of the maximum provided by the two stern thrusters. The bottom plot shows the estimated forward body referenced velocity and the results of two models fitted by least-squares. The first model was a discrete-time linear difference equation relating commanded force to velocity:

$$\dot{x}_{n+1} = \alpha \dot{x}_n + \beta u_n$$

where $\dot{x}_n$ is the velocity at time step n, and u_n is the commanded input force at time step n. The second model was a discrete-time nonlinear difference equation that approximates the effect of an abs-square drag law to first order:

$$\dot{x}_{n+1} = \alpha_1 \dot{x}_n + \alpha_2 \dot{x}_n |\dot{x}_n| + \beta u_n$$

Figure 4 shows that both models fit the data reasonably well, with the exception of the non-zero velocity at the start due to initial vehicle drift. The nonlinear fit was slightly better than the linear fit but not significantly. An explanation for this equivalence is that at such low speeds (<.1 m/s) damping effects are very low, and the vehicle behaves mostly as a pure mass. If the speed were to vary over a wider range, then the nonlinear model should provide a significantly better fit.

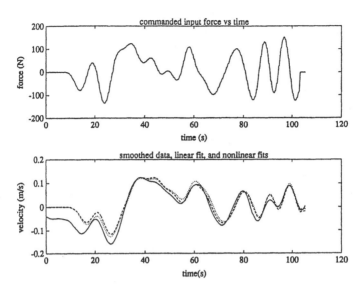

Figure 4. Results from the system identification of JASON's forward
dynamics. Linear and nonlinear models fit nearly equally well due to
the low range of speed in this test.

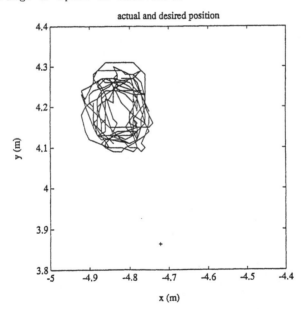

Figure 5. Performance of JASON in closed-loop hover. The "watch
circle" of about 10 cm radius is offset from the desired position by
the current. The limit cycle behaviour is likely due to a combination
of poor thruster performance at low amplitudes and time delays in the
multicomputer control system.

Figure 5 shows the results of a controller designed from the identified model. The controller utilized fixed position and velocity gains, with the full-state feedback provided by a steady-state Kalman filter. In this example, the control bandwidth was 0.5 radian/s. The vehicle holds station within a radius of approximately 10 cm (4 inches). The centre of the "watch circle" is offset from the desired position by approximately 30 cm.

The offset is caused by a moderate current (approx. 10 cm/s). An obvious solution to this problem is to add integral control action, although this could amplify the limit cycle behaviour. A better solution is to use adaptive techniques, such as those based on sliding control principals (Slotine, 1989 and Delonga, 1989), which are able to switch off the adaptation process after convergence.

While small in magnitude, the limit cycle behaviour is the most important aspect to improve in the context of coordinated vehicle-manipulator control. This behaviour seems to be caused by the interaction of several factors. These include delays, deadband, and hysteresis in the thrusters and pure delays in the system's computers and telemetry. While hydrodynamic nonlinearities are of primary concern in vehicles with a moderate or large speed range (Yoerger, 1985), at low speeds thruster deadband becomes a limiting factor in hover performance.

4.   JASON'S MANIPULATOR

JASON's manipulator was designed to be capable of regulating the interaction forces between the end effector and the environment from a basic design level. Such a capability greatly enhances manipulative capability either from a fixed base, in coordination with the vehicle, or with multiple manipulators. Details of the arm design can be found (DiPietro, 1988).

Rather than relying on force or torque sensing to permit good interactive behaviour, JASON's manipulator uses brushless servo motors coupled to the joints with cable reductions that feature zero backlash and low friction. The reducers are highly backdriveable. This permits the relatively high dynamic range of motor torque (controllable in feedforward to about 1%) to be reflected to the joint. Likewise, joint position can be estimated accurately from the resolvers built into the motors, which have a resolution of about 11 bits per motor revolution. The result is a manipulator capable of high performance control that utilizes no sensors other than those required to commutate the brushless motors.

The design of the motors, displacement sensors, and reducers allows high performance control to be implemented in a straightforward fashion. Sensing position at the actuator has fundamental stability advantages over sensing after the reduction at the joint. Good joint position or velocity control can be obtained using sensor information only from the motor due to the high performance of the reducers. Joint compliance can be actively controlled by adjusting the position gain in the joint servos. The manipulator is an excellent candidate for the application of a variety of schemes to control the interaction forces such as impedance control.

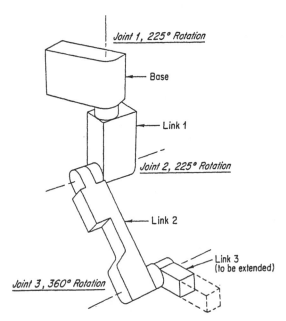

Figure 6. The first three links of JASON's manipulator. The
manipulator is designed using low friction, zero backlash cable
reductions that allow the joint torques and applied forces to be well
controlled through feed-forward torque control of the brushless dc
motors.

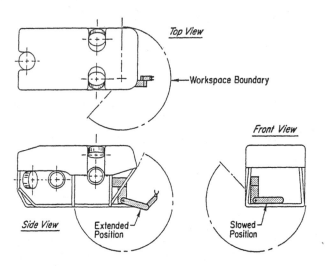

Figure 7. When mounted on JASON, the manipulator's wide range of joint
motion permits safe stowage inside the vehicle and excellent access in
front and under the vehicle.

The arm's shoulder and elbow joints are shown in figure 6. Each rotary joint is a modular element consisting of a sensorimotor (a servo motor with intrinsic high resolution position feedback) and a cable reduction. Each joint is flooded with ambient pressure oil and contains a low friction seal (John Crane). The reduction ratios are moderate, 30:1 in the shoulder and 13:1 in the elbow. Each joint has a wide range of rotation (240 degrees in the shoulder joints, 380 degrees in the elbow) which result in a large workspace and safe stowage, as shown in figure 7.

The choice of the sensorimotors (Seiberco, Inc.) was a crucial aspect of the design. They were chosen for many reasons, including the convenience of the built-in position resolvers, their high torque-to-weight ratio, low ripple torque, and the willingness of the manufacturer to repackage the control electronics to fit in JASON's 15.2 cm (6 inch) diameter pressure housings. Early in the program, the motors were run successfully in oil at a simulated depth of 6000 m in a pressure chamber.

The design of the cable reductions was the most challenging aspect of the project. A schematic of the shoulder joint reduction is shown in figure 8. The reduction must be sufficiently stiff to keep close correspondence between the motor position and the joint position and to prevent the introduction of low frequency dynamic modes. Keeping the design compact while respecting the required bend radii of the cables was also difficult. Providing the required pretensioning to eliminate backlash without introducing friction was another challenge. Careful design of the pulleys and terminations was required to prevent the cables from rubbing against each other or fatiguing.

Laboratory evaluations of the arm look very promising. The joints are straightforward to assemble, disassemble, and to pretension. The arm is very backdriveable and overall operation is very smooth. Figure 9 shows that the reductions are quite stiff. With the motor locked the arm deflects only about 1.5 mm under a 4 kg load. So far, the fatigue life of the reductions looks satisfactory, and over 100,000 cycles are expected.

Current development activities on the arm centre on the development of the control system, the design of a wrist, and overall integration with JASON. Field operation of the three joint arm with a modified commercially available end effector is expected in 1989.

5.   CONCLUSION

This paper summarizes two elements of the JASON control system. The first element examines the vehicle performance in hover and shows examples of an in-water system identification and closed-loop hover performance. While hover performance is very good (10 cm watch circle), this performance can be improved if the controller can deal with system delays and thruster nonlinearities. The second element describes the design of JASON's manipulator, which is designed to naturally implement force or impedance control schemes with minimal sensing. Laboratory evaluation of the manipulator confirms that the design goals can be reached in a full-ocean depth design.

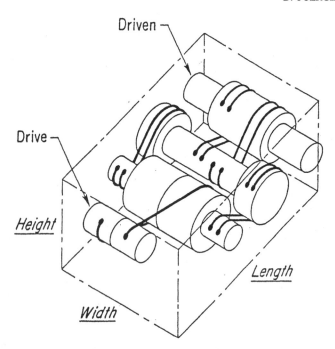

Figure 8.   Schematic of the three-state reductions used in the two
shoulder joints.

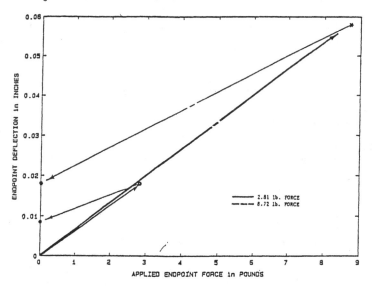

Figure 9.   This plot shows deflection of the arm as a function of the
applied load.   Despite the use of cable reductions, the arm is quite
stiff, deflecting only about 1.5 mm under a load of about 4 kg.   Upon
release of the load, the end of the arm returned to within 0.5 mm.

6.   ACKNOWLEDGEMENTS

James Newman was responsible for building the computer software
infrastructure for the JASON control system. Ralph Horber of Seiberco
Inc., (Braintree, MA) was of great help in the motor and controller
selection and procurement. The design and fabrication of JASON's arm
was due to the creative efforts of David DiPietro, with substantial
advice and guidance from Dr. Kenneth Salisbury of MIT. Hagen Schempf
assisted in the manipulator performance evaluation, made several design
refinements, and is now implementing the manipulator control system.
This work was supported by ONR contract N00014-86-C-0038, N00014-88-K-
2022, and ONR grant N00014-87-J-1111. This paper is Woods Hole
Oceanographic Institution contribution number 6959.

7.   REFERENCES

Delonga, D. (1989) "A control system design technique for
    nonlinear discrete-time systems", Ph.D. thesis, MIT-WHOI Joint
    Program for Oceanography and Ocean Engineering.

DiPietro, D.M. (1988) "Development of an actively compliant
    underwater manipulator", SM Thesis, MIT-WHOI Joint Program for
    Oceanography and Ocean Engineering.

SHARPS User Guide, Marine Telepresence Inc., Pocasset MA.

Slotine, J.J.E. (1989) "Sliding Controller Design for Nonlinear
    Systems", Int. J. Control, Vol. 40(2).

Von Alt, C. (1988) "Fiber Optics for Deep Ocean Systems", Proc.
    DOD Fiber Optics Conference.

Yoerger, D.R. (1985) "Robust trajectory control of underwater
    vehicles", IEEE J. Oceanic Eng., Vol. OE-10, No. 4.

SUBSEA WORK ENVIRONMENT FOR SUBMERSIBLES

J.L. MICHEL[1], J.F. DROGOU[1]
L. FLOURY[2]

ABSTRACT.   The manned deep sea submersibles CYANA and NAUTILE are used
increasingly as subsea work systems. This paper presents the
components of the subsea work environment of these vehicles:

    Telemanipulators and stocking
    Interfaces
    Tool and instrument panoply
    Shuttles

    An example of the installation of a subsea laboratory will be
given.
    The aim of future developments in telemanipulation will follow
thereafter.

1.   INTRODUCTION

The manned deep-sea submersibles CYANA (3000 m) and NAUTILE (6000 m)
are used increasingly for subsea work due to their capacities for
telemanipulation coupled with a work package environment.
    For scientific applications, a large range of equipment has been
developed for the sampling of rock sediment or water and for more
complex operations, such as in-situ measurements.
    A set of mechanical tool and rigging equipment has been highly
developed for rescue or salvage operations.
    The advantage of manned submersibles is their unique ability to
put equipment into operation at a precise location chosen in real-time
by the on-board observer, the pilot being able to react to
unpredictable situations.
    The manipulation capacities of the submersibles will be described
along with their tool and instrument panoply including subsea shuttles.

[1] IFREMER, Centre de Toulon, Zone Portuaire de Brégaillon, 83500 La
  Seyne Sur Mer, FRANCE.

[2] IFREMER, Centre de Brest, B.P. 70, 29263 Plouzane, FRANCE.

D. A. Ardus and M. A. Champ (eds.), Ocean Resources, Vol. II, 31–39.

2.   STOCKING AND MANIPULATION CAPACITIES OF THE NAUTILE

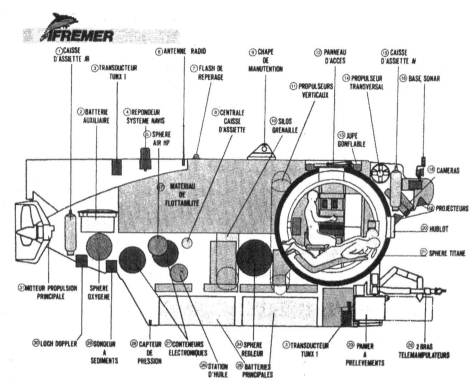

Figure 1.   Diagram of the NAUTILE.

The NAUTILE is fitted with a work arm/basket system mounted at the fore
in the direct field of vision of the pilots and passengers.
    The sample-basket has a holding capacity of 200 kg and a volume of
200 dm³.   In its retracted position, it diminishes only slightly the
field of vision of the lower portholes.   The basket can be jettisonned
for reasons of submersible security.
    Two telemanipulating arms are positioned at either side of the
basket.   One of these is the dexterous manipulator arm placed on the
port side directly in the field of vision of the pilot, the other is
the so-called "prehension" arm placed on the starboard side.   They are
entirely hydraulically operated, power being provided by the
submersible.   The hydraulic system is pressure compensated.   The
commands for each operating arm, of the "proportional" or "all or
nothing" type are situated on a joystick.   The pilot controls his
actions by direct vision through the sphere's portholes.
    In stocking position, the two arms are below the lower portholes,
hence out of the field of vision.
    This system was conceived with the aim of optimising the field of
action of the manipulators according to the field of vision and

stocking capabilities. A model of the front of the vehicle was created
using preliminary kinematic studies in order to verify the workings of
the system.

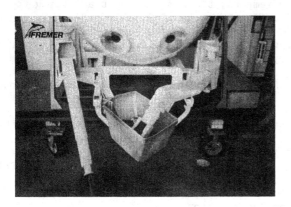

Figure 2. Model of the front of the NAUTILE: visibility, manipulation,
stocking.

2.1. Characteristics of the Dexterous Arm

Figure 3. Dexterous arm.

This arm possesses 6 degrees of freedom plus opening/closing mechanism
for the grip. The principal characteristics are as follows:

- 180° shaft rotation
- 90° shoulder articulation
- 120° elbow articulation
- 160° wrist rotation
- 120° claw articulation

-    continuous claw rotation
-    maximum length of 1.75 m

    During vertical movements, the arm can exert a lifting force of
70 daN, with a claw rotational couple of 9 daN.m and a grip force of
150 daN.
    Security regarding risks of hitches is ensured by means of an
entirely streamlined system with no visible hydraulic cable and a
jettison capacity.
    Different types of claw (parallel jaws, buckets) can be mounted
prior to the dive.

2.2. Characteristics of the Prehension Arm

The arm possesses 4 degrees of freedom plus an opening/closing
mechanism for the grip. The principal characteristics are as follows:

-    180° shaft rotation
-    15° - 30° shoulder articulation
-    300 mm extension
-    360° claw rotation
-    maximum length of 1.75 m

    The arm can exert stresses of an equal value to those of the
dexterous arm and can also be jettisoned.

3.   STOCKING AND MANIPULATION CAPACITIES OF THE CYANA

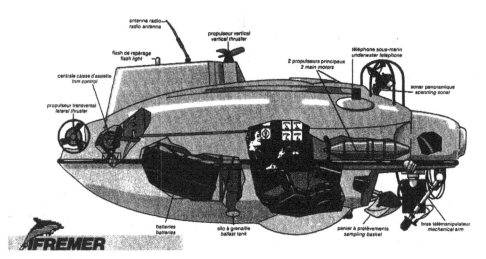

Figure 4.   Diagram of the CYANA

The telemanipulating arm which is positioned underneath the cap possesses 4 degrees of freedom, plus an opening/closing device for the claw. Its maximal extension is of 1 m, and the manipulating arm can exert a lift force of 20 kg in the vertical plane. It is hydraulically operated and the "all or nothing" type commands are situated on a joystick.

In stocking position, the arm is entirely out of the field of vision. During manipulation, the pilot sees only the end of the arm and operates using its direct vision through the porthole.

The arm permits the carriage of samples and equipment whose in-air weight does not exceed 20 kg. Manipulation is carried out for the most part "in flight", taking advantage of the supplementary degrees of freedom of this highly manoeuverable submersible (displacement, giration or pitch).

The port-side retractable basket can hold up to 30 kg of samples.

4.   INTERFACES

The capacities for manipulation permit the direct taking of samples of objects such as stones, nodules and the reclamation of small lost items.

In order to reply to the wishes expressed by the scientific community as well as the requirements linked to submersible security, the telemanipulation system is equipped with a large range of tools and instruments.

The interfaces between the vehicle equipped with manipulating arms and the tools have been the subject of numerous analyses.

Standard tool/manipulating arm interfaces have been tested:

-    for prehension

     *   The T-bar allows for precise positioning and a rigid hold of
          the tool. The arm should, however, be relatively dexterous
          in order to seize the T.

     *   The Ball facilitates grasping by a hardy arm when precision
          and rigidity are not essential.

     *   The Plate adapts well to parallel-jaw claws, ensuring a rigid
          grip.

-    for mechanical commands, the manipulating arm can:

     *   exert a direct translatory force

     *   exert a direct grip force

     *   seize directly a tool and exert force indirectly by the
          addition of a supplementary hydraulic actuator.

For the other active or instrument-equipped tools, mechanical, electrical or hydraulic interfaces must be provided.

The designer must equally take into account the safety requirement for the submersible: all the tools must effectively be able to be jettisonned at all times.

An electro-hydraulic connecting/disconnecting module was created for the former, as well as a pyrotechnic interface-jettison module for the latter.

5.   EQUIPMENT AND LIGHT INSTRUMENTS

Using the interfaces defined above, the following equipment is put into operation by the submersibles:

-    water-sampling bottles
-    probes (temperature and resistivity measurement)
-    sediment corer (Cahet, 1986; Sibnet, 1985)
-    hydraulic hammer for rock
-    rock corer
-    cylindrical vacuum sucker
-    vane test
-    thermostatically controlled enclosure
-    marking bell (Van Praet, 1985)
-    marking box
-    tracer injector (formalin).

For operations of salvage or reclamation of lost items, specific tools have been developed:

-    wire cutters for cutting different sorts of metal or synthetic cables;

-    prehension tools: hooks and grapnels.

6.   INSTALLATION OF SUBSEA LABORATORIES

By combining manipulation, equipment and stocking, the team aboard the submersible can carry out complex operations.

In this way, during the Kaiko campaign, a Japanese geophysical laboratory was installed by the NAUTILE in 3930 metres depth under Erimo, a subsea volcano at the junction of the trench of the Kouriles and Japan. A seismometer and two tiltometers were set up at around one hundred metres distance.

The two tiltometers were cemented to the sea-bed during a 17-hour diving mission by the NAUTILE.

Later, information gathered from this equipment is then collected periodically from surface level.

CHAPTER 4

7.   EXTENSION OF THE WORKING ENVIRONMENT: THE SHUTTLE CONCEPT

The holding capacity of submersibles is limited. This is particularly important in the case of CYANA which has a restricted on-board volume and weight limit (a few dm³ and several tens of kg).

    An increase in the direct capacities of the vehicle would necessitate important modifications.

    The concept of the submersible shuttle has been developed to respond to the need for an increase in the number of samples taken and the use of a wider panoply of equipment.

    The concept has the following advantages:

-    it allows for a quick response to new needs;

-    stocking can be increased and equipment of a large volume can be put into operation;

-    the submersible is not obliged to devote a dive entirely to one task.

Figure 5.   Free subsea shuttle for delivering instruments to the submersible.

Several types of submersible-implemented shuttle have been developed:

-   autonomous shuttles destined for the transport of instruments or samples;

-   the NADIA type specific-tool shuttles (Legrand, 1988).

Let us describe more precisely the autonomous shuttle which is a "free" type piece of equipment used for the transport of instruments put into operation by the submersibles.
The position on the seabed between the shuttle and the submersible is ensured by means of an acoustic long baseline system.
The shuttle can be displaced by the submersible in order that it may be placed near to the work zone.
The instruments are placed in two 1 m^2 section drawers of 0.6 m in height, representing a usable volume of 1.2 m^3 (maximum weight capacity: 60 kg in water).
The drawers are opened by means of a telemanipulating arm.
The shuttle integrates an acoustic remote control for the realisation of the following functions with the remit:

-   to drop the descent weight;

-   to fill the ballast for increasing the weight of the shuttle after a possible displacement;

-   to drop the ascent weight.

These three sequences can also be commanded by a telemanipulating arm.

8.   FUTURE DEVELOPMENTS

We have described the components of a work environment including:

-   an integrated on-board visibility, manipulation and stocking system;

-   hydraulic, mechanical and electrical interface mechanisms;

-   an equipment and instrument panoply;

-   free shuttles.

On this basis, a large applicational field is, or becomes possible.
Operation time remains, however, relatively long. For example, the simple procedure of the collection of a sample-bottle and its stocking requires several minutes. Present limitations are situated at a command level. Important progress can be expected from more ergonomic or automated commands for repetitive functions.

Operational reliability can also be improved by the limitation of
the extreme stress constraints inflicted upon telemanipulators operated
by highly inert vehicles and by the determination of no-go zones.

For this purpose, and with a view to replying to industrial
requirements, IFREMER is developing a test-bed system based on computer
aided teleoperation (Leveque, 1988). Bilateral force feedback
telemanipulation will thus be tested with computer assistance both in
teleoperating and robotic modes.

Access to new fields of work makes possible the simplification of
procedures and equipment. An increase in operational reliability and
above all considerable time-savings are to be expected from an advanced
system of telemanipulation.

9.   REFERENCES

Cahet, G., Sibuet, M. (1986) 'Activité biologique en domaine
    profond: transformations biochimiques in situ de composés
    organiques marques au carbone 14 à l'interface eau-sediment par
    2000 metres de profondeur dans le Golfe de Gascogne', Mar. Biol.
    90, pp.307-315.

Sibuet, M., Floury, L., Cahet, G. (1985) 'In situ deep-sea
    biological activity at the sediment-water interface by
    experiments with a submersible: (1) Submersible instrumentations
    developed for the program BIOCYAN. (2) In situ biotransformation
    of labelled organic matter at the sediment-water interface in
    the Bay of Biscay at 2000 m depth', 4th Symposium on Deep-Sea
    Biology, June 23-25, 1985, Hamburg RFA.

Van Praet, M., Sibuet, M. (1985) 'Métabolisme in situ du benthos
    profond: éssais d'une enceinte sur les Cnidaires', 2eme journées
    d'étude de la plongée scientifique, Nice, 15-16 mars, 1984,
    Bull. Inst. Oceanogr., numéro spécial 4, pp.185-190.

Legrand, J. et al (1988) 'NADIA: wireline re-entry in deep-sea
    boreholes', Proc. Symp. OCEANS 88.

Leveque, C. et al (1988) 'Advanced remote manipulation test bed
    for subsea applications', 2nd Int. Symp. on Subsea Robotics,
    Tokyo, November 1988.

THE JAPANESE MANGANESE NODULE MINING SYSTEM

YUJI KAJITANI
Metal Mining Agency of Japan
Tokyo
JAPAN.

1.   THE RESEARCH AND DEVELOPMENT PROGRAM FOR THE MINING SYSTEM

Research and development for the manganese nodule mining system was
initiated in Japan by the Deep Ocean Minerals Association (DOMA) in the
1970s. They were searching mainly for an efficient mining concept.
The members consisted of engineers from the non-ferrous metal,
machinery and shipbuilding industries and of staff of the Ministry of
International Trade and Industry (MITI).
      On the basis of these preliminary studies, the Agency of
Industrial Science and Technology (AIST), part of MITI, decided to
develop the mining concept and established the following research and
development program:

-    R & D period: nine years from 1981 until 1989

-    R & D budget: twenty billion Japanese Yen

-    Basic concept of the mining system: hydraulic dredge with towed
     collector

MITI also decided that this program should be sponsored and directed by
AIST.
      Basic research was to be carried out by the National Research
Institute for pollution and Resources (NRIPR), while engineering work
was to be undertaken by the Technology Research Association (TRA) of
the Manganese Nodule Mining System which was organized for this
program. The TRA consists of twenty companies from such industries as
non-ferrous metal mining, shipbuilding, machinery, electrical products
and shipping. The Metal Mining Agency of Japan is a member of TRA.
      The TRA produced a concept design in 1981, followed by R & D of
elemental technology for the collector, the manganese nodule lift, the
machinery handling system, and the measurement and control system.
      Necessary scale models were manufactured and tested in model
basins and in the sea. Many simulation calculations were also
required. This stage of the R & D was completed in 1985.

41

D. A. Ardus and M. A. Champ (eds.), Ocean Resources, Vol. II, 41–44.
© 1990 *Kluwer Academic Publishers. Printed in the Netherlands.*

The basic design of the mining system was established simultaneously, taking into account the results of the R & D carried out regarding elemental technology. This stage too was completed in 1985. In our program, a pilot study involving the raising of actual manganese nodules will be carried out in the mining area during the last year of the program. Therefore, detailed design was initiated in 1986 in order to manufacture the necessary equipment. This detailed design stage is still in progress since designs for some pieces of equipment are dependent upon those for others. However, the manufacture of some items has already started.

Although the R & D program has made good progress, there have been delays totalling a few years. There are two reasons for these delays. The first is technical difficulties. We have met many kinds of obstacles and overcome almost all of them. However, there are still a few, to be dealt with. The second reason is budgetary restrictions. Commercial production has been delayed, so it is difficult to obtain immediate funds for our program. As a result, our pilot study which was scheduled for 1989 will be delayed by a few years.

2.   AN OUTLINE OF THE JAPANESE MANGANESE NODULE MINING SYSTEM

The mining system being developed in this program employs a hydraulic dredging method. Manganese nodules are collected by a towed vehicle on the ocean floor and lifted to a surface ship by the hydraulic lift system using hydraulic pump and/or air lift pump through a long pipe of several thousand metres.

This mining system consists of the following five systems.

(1)   Total System
       To forecast performance requirements for a future system of
       commercial recovery of manganese nodule, to develop experimental
       recovery system including a mining ship, to conduct experiments in
       each stage of development, to verify operability of the
       experimental system through land and ocean mining tests, to
       collect data for designing a future system of commercial
       production, and to plan experiments efficiently to develop
       production engineering. Each of the following systems are
       coordinated and integrated throughout performance of these tasks.

(2)   Collector System
       To develop a towed collector. The collector collects manganese
       nodules efficiently spread in a large quantity in the deep seabed,
       discards the dump, and feeds the next lift system with the slurry.

(3)   Lift System

       (a)   Pump lift equipment
             To develop a high-head submerged pumping system of slurry,
             and a submerged motor resistant to water pressure. This
             system supports a lift system under research and development
             to feed the mining ship with the collected manganese nodules.

(b) Air lift equipment
     To develop a lift system to feed the mining ship with the
     collected manganese nodules using compressed air blown into
     the lift pipe. This system takes advantage of the difference
     in gravity between air and sea water. Also, to develop an
     accessory to separate gas, liquid and solid.

(c) Lifting pipe equipment
     This system is an artery to feed the mining ship with the
     slurry. It is used as the cable to tow the collector.
     Research and development of this system involves development
     of high-tension steel pipe, high-tension rubber hose to
     connect collector with lift system, a fairing to reduce drag
     incidental to towing, and device to protect pipes from being
     blocked.

(d) Machinery handling system
     To perform research and development on a hoist to carry
     submerged equipment firmly, safely and fast to and from the
     mining ship.

(e) Measurement and control system
     This is under research and development to coordinate and
     integrate the interfering requirements on measurement and
     control by total system and each operating system, and
     consequently to perform ocean mining tests with ease and
     efficiency. It involves engineering development of complex
     cable for power transmission, optical fibre, and connector
     resistant to high tension and hydraulic pressure equivalent
     to 500 atmospheric pressure.

3. SOME ITEMS TO BE SOLVED BEFORE COMMERCIAL MANGANESE NODULE MINING
    CAN BE ATTEMPTED

Our R & D program aims to carry out a pilot study in the Pacific Ocean.
We believe that the study will be successful. However, even if the
results are good, there are still some technical and economic problems
remaining before commercial manganese nodule mining can be attempted.
I shall deal only with the technical problems here. Firstly, more
detailed information will be needed regarding the sea bed where the
manganese nodules are located. Our mining system utilizes a towed
collector. Therefore, undulations, particularly any steep hills or
precipices, would create problems. For our system to operate,
relatively flat areas have to be selected. This means that knowledge
of the precise configuration of the sea bed is necessary. It follows
that we must ascertain the ship's position with a high degree of
accuracy and have highly sensitive depth gauging equipment in order to
be able to map the sea bed. Moreover, we have to develop a ship which
is capable of maintaining a designated course with low speed and high
power. Secondly, reliable equipment for long-term commercial
production has yet to be developed. Taking the pump as an example, sea

water containing nodules must pass through the pump so rapidly that frictional damage is bound to occur. The elements of the pump must be strong enough to withstand this friction. Similar difficulties would be experienced with other equipment. For long term commercial production, many pieces of equipment need to be submerged in sea water. Friction, errosion and fatigue are unavoidable. These common factors would also affect the flexible hoses and steel pipes. Thirdly, extension of the pilot study to the scale necessary for commercial production must be carried out with care. Theory alone will not suffice to determine how to apply the results of the study to commercial production. Finally, there are unknown factors in the ocean, especially in the deep ocean. Even though a great deal of research has been carried out all over the world, we still do not know everything about conditions around the bottom of the deep ocean. To this extent, we must also admit that there is the possibility that the pilot study will be a failure.

Those, then, are the four technical obstacles to the commercial mining of manganese nodules using our method. Economically, of course, commercial production is still far away. However, we shall continue to strive to develop the necessary technology to extend the frontiers of deep sea exploitation.

SEA BED SAMPLING WITH AN EXPENDABLE ACOUSTIC PENETROMETER SYSTEM

REG CYR
Sonatech Inc.
Santa Barbara
California
U.S.A.

ABSTRACT. This paper describes an expendable, acoustic, dynamic penetrometer system that is used to determine the undrained shear strength of ocean bottom sediments to a depth of approximately 9 m in water depths to 6 km. Doppler frequency shifts associated with changing velocities during bottom penetration are measured relative to a highly stable sound source which is attached to a slender weighted body. The varying rates of deceleration are correlated by a topside algorithm into forces which characterize the undrained shear strength of sediment layers down to the depth of penetration. In addition to a physical description of the hardware developed for this purpose, the method for determining soil strength and correlation to unconsolidated, undrained triaxial shear and mini-vane tests on samples recovered with a Shelby tube sampler are discussed.

Though initially designed for site selection and design of embedment anchors, the geotechnical properties that are measured make it a useful tool for a much broader applications spectrum, such as large scale EEZ surveys. In addition, though expendable, it offers a cost-effective alternative to augment and, in some cases, replace coring surveys. Expansion concepts into "in situ" soil sound velocity and acoustic propagation losses are suggested.

1. THEORY OF OPERATION

The CW sound source provides a velocity vector which indicates probe terminal velocity and impact velocity change (deceleration) during penetration, as illustrated in Figure 1.

The doppler equation can be expressed as:

$$V_s = V_f = \frac{F - F_1}{F_1}$$

where:

V_s = Velocity of the source

V_f = Velocity of sound in the surrounding medium

F = Basic oscillator transmitter frequency

F_1 = Perceived frequency

D. A. Ardus and M. A. Champ (eds.), Ocean Resources, Vol. II, 45–56.
© 1990 *Kluwer Academic Publishers. Printed in the Netherlands.*

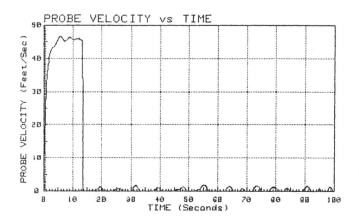

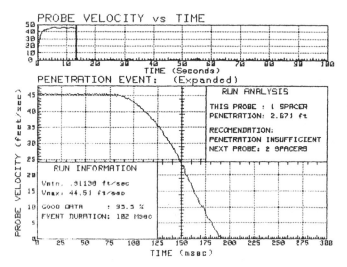

Figure 1.   Typical EDP graphic/digital display.

      Parameters which effect soil failure include frontal bearing
resistance, side resistance, buoyancy, inertial or drag forces, and
added mass. True, 1975, developed the following equation to convert
penetrometer motion to soil strength:

$$M'V' \frac{dv}{dz} = F_D + W_b - F_{BE} - F_{AD} - F_H$$

where:

M' = Penetrometer effective mass

$\dfrac{dv}{dz}$ = Differential penetrometer velocity with soil depth

F_D = External driving force (0 for free fall body)

W_b = Buoyant weight of penetrator

F_{BE} = Bearing component force

     $= S_e^{\cdot} \, (S_u \, N_c \, A_f)$ (True, 1975)

where:

$S_e^{\cdot}$ = Soil strength strain rate factor

S_u = Soil undrained shear strength

N_c = Bearing capacity factor

A_f = Penetrator frontal area

F_{AD} = Side adhesion force

     $= S_e^{\cdot} \, \left(\dfrac{S_u \, A_s \, \delta}{S_t} \right)$ (True, 1975)

where:

A_s = Penetrometer side area

S_t = Soil sensitivity

δ = Side adhesion factor (intentionally neglected by Beard, 1977).

F_h = Inertial or drag force

     $= 1/2p \; C_d \, A_f \, V^2$

where:

p   = Fluid or soil mass density

C_d = Drag coefficient

    True formulated the strain rate factor to the penetrometer velocity, diameter and the undrained shear strength:

$$S_e^{\cdot} = S_e^{\cdot} \, max \; \left(1 + \sqrt{\dfrac{C_e^{\cdot} \, v}{S_u \, t}} + 0.6 \right)$$

where:

$C_e^{\cdot}$ = Empirical strain rate

        Coefficient, 1900 Pa-sec (40 lbs - s/ft^2)

As for $S_e^{\cdot}$ max, the following maximum values have been suggested:

SOURCE	$S_e^{\cdot}$ max
True	4
Prevost	1.5
Beard	2

Sediment sound velocity is extrapolated from Hamilton and Bachman (Refs., 1982). Assumed values of V_s as well as Bulk Wet Density are given below: (Beard, 1983).

CONDITION	ASSUMED V_s IN SEDIMENT	ASSUMED BULK WET DENSITY
Penetration $\geq$ 7.5 m	V_s water $-$ 1.5%	1440 Kg/m³
7.5 m > Penetration > 4.5 m	$= V_s$ water	1600 Kg/m³
Penetration < 4.5 m	1.05 V_s water	1760 Kg/m³

V_s in sediment assumed   0.5 m/s/m   Parameter studies
                                                   showed these values
                                                   held errors to <1%

Assumed sediment sensitives based on N.C.E.L.´s field data (Lee, 1973).

SEDIMENT	SENSITIVITY
Pelagic Clay	3
Calcareous Clay	4-6
Terrigenous Clay	3
Clayey Silt	2
Silty Sand/Sandy Silt	2

Final data reduction is done incremently in an iterative fashion. A computer program has been developed to edit, calibrate and smooth the acquired data; a cubic-spline curve fit is used to calculate velocity and deceleration. The velocity depth profile is used to calculate the soil strength profile and the deceleration depth curve the shape of the strength profile.

2.   SYSTEM HARDWARE DESCRIPTION

Figure 2 is a photograph of the PX-010 probe and the topside processor. Specifications for the major components are listed below:

SYSTEM HARDWARE DESCRIPTION

Figure 2 is a photograph of the PX-010 probe and the topside processor. Specifications for the major components are listed below:

	MODELS			
	S1070	S1156A	S1156B	PX-010
Frequency	12kHz $\pm$ 0.01%	12kHz $\pm$ 0.01%	12kHz $\pm$ 0.01%	60kHz nominal
Source Level	+190dB $\pm$ 2dB ref 1 uPa @ 1 meter	+180db $\pm$ 5dB ref 1 uPa @ 1 meter	+190 nominal ref 1 uPa @ 1 meter	162dB ref 1 uPa @ 1 meter (nom)
Maximum Depth	6000 meters	6000 meters	6000 meters	1200 meters*
Power Supply	Sealed Lead Acid	Manganese Dioxide	Manganese Dioxide	Non-recharge-able Alkaline
Life	10 min. design (Signals detected for 1 hr.)	Clock Shutdown 11 min.	Clock Shutdown 11 min/3 min	104 sec. max. (after probe launch)
Trans-ducer Type	Longitudinal Vibrator	Cavity Resonator	Longitudinal Vibrator	Piston
Beam-width	$\pm$45° (-3dB)	Hemispherical	$\pm$25° (-3dB)	$\pm$25° (-3dB)
Probe Size	0.46M x 90mm D	0.3M x 90mm D	0.3M x 90mm D	0.96M x 90mm D
Body Size	2.45M x 90mm D	2M x 90mm D	2M x 90mm D	0.96M x 90mm D
Body Weight	161Kg	134Kg	134Kg	27.2Kg

*Deeper models on request.

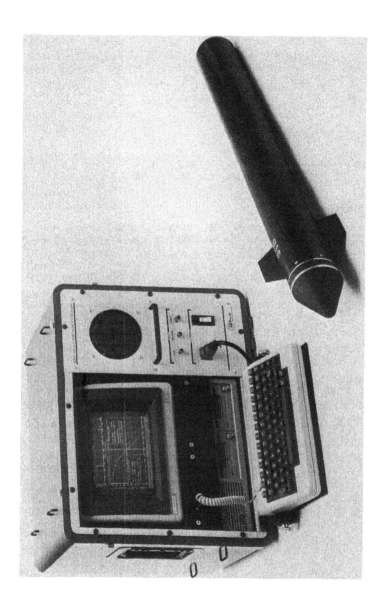

Figure 2. PR010 doppler penetrometer receiver and PX010 expendable probe.

3.   FIELD TEST

Tests were performed in pelagic clay, calcerous ooze and sandy soils to
silty clays (terrigenous deposits). Test water depths ranged from 30
to 5490 m. Table 1 documents the pertinent facts about the launch site
areas.

4.   SAMPLE TEST SITES

Figure 3 illustrates the results of two penetrometer drops at the top
of the ridge. A gravity core was taken and tested aboard ship by mini-
vane to determine strength profiles. The penetrometer drops appear
consistent with each other and are in good agreement with the mini-vane
measurements.

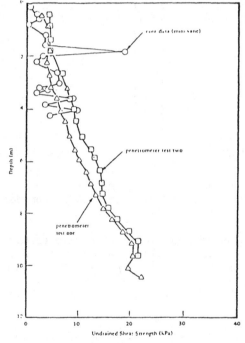

Figure 3.   Pelagic Clay, comparative undrained shear strength data,
Gulf Stream Outer Ridge, water depth 4770 m.

     Figure 4 depicts the results of two penetrometer drops compared to
triaxial tests performed on a piston core sample. The penetrometer
drops are consistent with each other and in good agreement with the
reference.
     The results outlined in Figure 5 were acquired in relatively
shallow water, off the coast of California, in low plasticity, silty
clay. The reference measurements are unconsolidated, undrained
triaxial shear and mini-vane tests on samples recovered with a Shelby
tube sampler. Again, basic agreement is seen, although the reference

measurements may be in error somewhat as the sampler was driven with a
hammer.

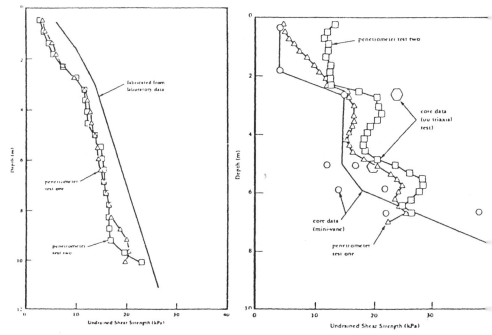

Figure 4. Calcareous Ooze,
comparative undrained shear
strength data, Carribean Sea,
water depth 3690 m

Figure 5. Silty Clay, comparative
undrained shear strength data, San
Pedro Bay, water depth 165 m.

Figure 6 depicts an adaption of the acoustic bottom penetrator to
provide a measure of the sound velocity and absorption of the sub-
bottom strata, which in turn may provide information on mean grain size
or porosity.

5. CONCLUSION

The acoustic doppler penetrometer has met all of its design goals. The
present configuration has demonstrated its ability to provide a
reasonable estimate of the undrained shear strength profile down to its
penetration depth. Additionally, the simple penetration depth itself
provides a good indication of the average undrained shear strength.
The doppler penetrometer can be made in many configurations depending
on the type and depth of measurement desired and can be modified to
provide a measurement of other parameters of interest such as sound
velocity and attenuation.

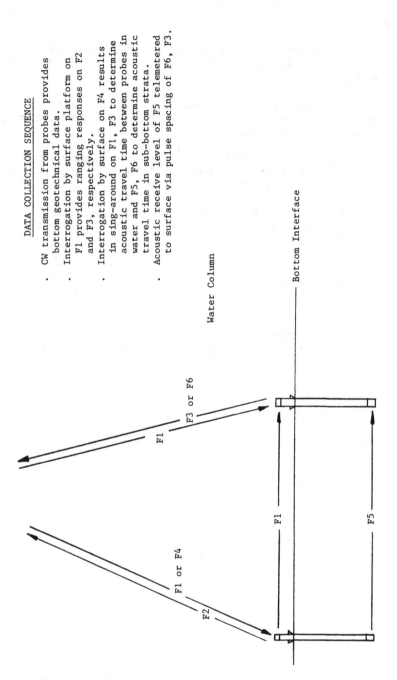

DATA COLLECTION SEQUENCE

- CW transmission from probes provides bottom geotechnical data.
- Interrogation by surface platform on F1 provides ranging responses on F2 and F3, respectively.
- Interrogation by surface on F4 results in sing-around on F1, F3 to determine acoustic travel time between probes in water and F5, F6 to determine acoustic travel time in sub-bottom strata.
- Acoustic receive level of F5 telemetered to surface via pulse spacing of F6, F3.

Figure 6.   Adaptation of acoustic penetrator to measure sub-bottom acoustic velocity and attenuation.

6.   REFERENCES

Beard, R.M. (1977) 'Expendable doppler penetrometer: A perforamnce
    evaluation`, Civil Engineering Laboratory, Technical Report R-855,
    Port Hueneme, CA.

Beard, R.M. (1983) 'Expendable doppler penetrometer for deep ocean
    sediment strength measurements`, Naval Civil Engineering Laboratory,
    Technical Report R-905, Port Hueneme, CA.

Hamilton, E.L. and Bachman, R.T. (1982) 'Sound velocity and related
    properties of marine sediments`, Journal of the Acoustic Society of
    America, Vol. 72, NO. 6, pp. 1891-1904.

Lee, H.J. (1973) 'In-situ strength of seafloor soil determined from
    tests on partially disturbed cores`, Naval Civil Engineering
    Laboratory, Technical Note N-1295, Port Hueneme, CA.

Prevost, J.J. (1976) 'Undrained stress-strain-time behavior of clays`,
    Proceedings of the American Society of Civil Engineers, JOurnal of
    the Geotechnical Engineering Division, Vol. 102, No. GT12,
    pp. 1245-1259.

True, D.G. (1975) 'Penetration of projectiles into seafloor soils`,
    Civil Engineering Laboratory, Technical Report R-822, Port Hueneme,
    CA.

TABLE 1. Site Descriptions

Date	Site No.	General Location	Geographic Coordinates		Water Depth		Soil Description
			Latitude (N)	Longitude (W)	Meters	Feet	
Aug 1976	I	Santa Barbara Channel	34°17.2'	119°42.8'	180	600	Nonuniform terrigenous deposit. Firm sandy clayey silt to 1 m; clayey silt below 1 m (ML-MH).
Aug 1976	II	Santa Barbara Channel	34°16.5'	119°	370	1,200	Soft, uniform terrigenous plastic clayey silt (MH).
Aug 1976	III	Santa Cruz Basin	33°51'	119°41'	1,830	6,000	Soft, terrigenous silty clay with occasional sand lens. (MH).
Aug 1976	IV	Santa Monica Basin	33°43.5'	119°05'	880	2,890	Uncored terrigenous soil. Thought to be cohesive based on soil found on anchor fluke.
Dec 1976	V	N.E. Pacific Ocean	21°32.6'	143°38.6'	5,430	17,800	Pelagic clay.
Dec 1976	VI	San Diego Trough	32°33.5'	117°29'	1,230	4,040	Firm, terrigenous clayey silt (MH); some sand lenses.
Sep 1977	VII	Blake Plateau	27°59.8'	77°10.8'	1,130	3,700	Calcareous ooze, cohesionless. Coarse grained, very sensitive, 80% carbonate content.
Sep 1977	VIII	Nares Abyssal Plain	21°01'	66°24'	5,490	18,000	Pelagic clay, soft manganese nodules.
Jun 1978	IX	Sohm Abyssal Plain	34°43'	61°24'	4,570	14,990	Pelagic clay, soft.
Aug 1978	X	Santa Cruz Basin	33°51'	119°41'	1,700	5,580	Terrigenous silty clay with occasional sand lenses.
Oct 1978	XI	Gulf Stream Outer Ridge	34°49.9'	65°50.5'	4,770	15,650	Pelagic clay.
Oct 1978	XII	Gulf Stream Outer Ridge	36°04.7'	66°13.4'	4,835	15,860	Pelagic clay.
Oct 1978	XIII	Gulf Stream Outer Ridge	36°09.8'	66°17.2'	4,908	16,090	Pelagic clay.
Dec 1978	XIV	San Pedro Bay	33°35'	118°07.7'	75	250	Terrigenous sandy silt (ML) to clayey silt to 4.5 m, silty clay below.

continued

TABLE 1. Continued

Date	Site No.	General Location	Geographic Coordinates		Water Depth		Soil Description
			Latitude (N)	Longitude (W)	Meters	Feet	
Dec 1978	XV	San Pedro Bay	33°34.4'	118°07.2'	165	540	Terrigenous soft to medium silt (CL) to 6 m, stiff clayey silt below 6 m.
Dec 1978	XVI	San Pedro Bay	33°33.9'	118°07'	215	700	Terrigenous very soft to medium stiff clayey silt (ML-CL).
Feb 1979	XVII	Santa Barbara Channel	34°11'	119°27.8'	95	310	Terrigenous soft to firm silty clay (CL) to 3 m, medium dense sands and silts 3 to 6 m.
Feb 1979	XVIII	Santa Barbara Channel	34°19.9'	119°33.4'	55	180	Terrigenous sandy silt (ML), hard, low plasticity.
Jun 1979	XIX	Puget Sound	48°03'	122°46'	30	100	Soft silty organic clay (OH) of medium to high plasticity.
Oct 1979	XX	Caribbean Sea	16°58.9'	74°01.2'	3,730	12,230	Calcareous ooze, 50% carbonate carbon, 60% clay sized, sensitive, firm.
Oct 1979	XXI	Caribbean Sea	17°0.9'	79°30.5'	1,130	3,710	Calcareous ooze, 74% carbonate carbon, 60% clay sized, very sensitive, soft.
Oct 1979	XXII	Caribbean Sea	15°34.9'	78°22.8'	700	2,300	Calcareous ooze, 60% carbonate carbon, 55% clay sized, moderately sensitive, soft.
Oct 1979	XXIII	Caribbean Sea	14°46.4'	78°3.7'	1,880	6,170	Calcareous ooze, 51% carbonate carbon, 35% clay sized, very soft.
Oct 1979	XXIV	Caribbean Sea	13°16.2'	78°2.6'	3,690	12,000	Calcareous ooze, 44% carbonate carbon, sandy clay, very soft.

A SITE-SPECIFIC SAMPLING SYSTEM FOR EEZ HARD MINERAL DEPOSITS

JOHN R. TOTH and CRAIG A. AMERIGIAN
Analytical Services Company
Carlsbad
California
U.S.A.

ABSTRACT. Accurate evaluation of Exclusive Economic Zone (EEZ) hard mineral deposits requires survey and sampling techniques more advanced than those currently in use. In response to this need, a tethered, multiple coring system for the recovery of site-specific samples of manganese crust has been designed, fabricated, and tested at sea. This system consists of an array of thirty core guns, bottom triggering and sequencing electronics, an acoustic telemetry system, and a 35 mm camera system. It can be operated to depths up to 5000 m from typical oceanographic research vessels using a standard mechanical deep sea wire.
    The coring system was deployed in March 1987 from the French research vessel Jean Charcot, sampling crust deposits in the Tuamotu archipelago. Based on these successful sea trials, plans have begun for a number of future developments:

(1)   refinement of the current sampler for easier rigging on deck and to enable greater core penetration,

(2)   modification of core barrels to recover other consolidated substrates for site-specific sampling of glassy and weathered basalt and massive sulphides,

(3)   interfacing of the system with a conducting cable and addition of real-time video and remote firing to enable interactive selection of sampling sites, and

(4)   design of a small, untethered sampler that will recover a single sample and return to the surface, similar in concept to the free-fall grab samplers used in exploration for manganese nodules.

1.   INTRODUCTION

The resource potential of deep water marine minerals has been brought into focus with the establishment of 200-mile offshore Exclusive Economic Zone (EEZ's) by the world's coastal and island nations.

57

D. A. Ardus and M. A. Champ (eds.), Ocean Resources, Vol. II, 57–68.
© 1990 *Kluwer Academic Publishers. Printed in the Netherlands.*

Programs are currently underway in a number of nations for evaluating
the potential deep ocean mineral resources that exist within their 200
mile boundaries. The deposits of greatest concern are cobalt-rich
manganese crusts, occurring on oceanic seamounts, and zinc, copper and
possible precious metal rich polymetallic sulphides, occurring on
active oceanic centres and seamounts.
    The United States efforts have primarily centered on

(1) detailed surveying of the EEZ with the GLORIA sidescan sonar and
     swath bathymetric mapping systems,

(2) evaluation of manganese crust deposits within the Hawaiian
     archipelago and assessment of the environmental impact of crust
     mining,

(3) surveys of the hydrothermal sulphide deposits on the Juan de Fuca
     Ridge and

(4) establishment of a regulatory framework for mining claims.

Extensive survey operations have also been conducted by the French for
crust deposits in the Tuamotu archipelago and the Canadians for
sulphide deposits on the Juan de Fuca Ridge. In addition Japan, Korea,
Australia, Germany and the Soviet Union reportedly have operations
underway or planned either independently or in partnership with Pacific
Island nations for survey of manganese crust and other deposits.
    The primary tools for evaluating EEZ hard mineral deposits have
been sidescan sonar and swath bathymetric mapping systems for areal
reconnaissance, near bottom photography or video imaging, and dredge
sampling. The acoustic reconnaissance operations utilize well
established and proven hardware. Little work, however, has been done
in the development of seafloor sampling and near bottom survey
instrumentation for site-specific sampling and determination of
seafloor characteristics at potential mine sites.
    Sampling technology in particular has seen little development.
The design and navigation of dredges have been improved in recent years
but these devices still are limited to recovering bulk material from a
relatively large area and are highly biased toward recovery of loose
rubble and material on local highs. As a result, recent workshops and
reviews of EEZ research (Lockwood and Hill, 1985; Yuen, 1986) have
recommended the development of seafloor systems, especially for
recovery of hard mineral samples, as an important priority.
    Lack of effective, site specific-sampling techniques has made
detailed study of mineral deposits and other consolidated deposits
difficult. Such systems are needed for collection of

(1) statistically significant numbers of hard mineral samples to allow
     resource assessment,

(2) mineral and underlying substrate samples for mining engineering
     evaluation, and

(3)   samples of presumed hard rock occurrences in order to verify
      geological interpretations made from remotely collected acoustic
      data.

2.   DEVELOPMENT OF ASC CRUST CORER

Since 1984 Analytical Services Company (ASC) has been involved in the
design and testing of systems for recovering discrete samples of
manganese crust and other seafloor consolidated deposits. These
systems utilize percussion coring technology, similar to that used in
sidewall coring devices by the oil services industry to recover samples
from the walls of drill holes. With these systems samples are taken by
firing tethered, hollow core barrels, using small explosive charges,
into the sampling substrate. The ASC development program has
progressed from the design and land-based testing of individual coring
guns to the fabrication and testing at sea of two multiple-core devices
for the recovery of manganese crust samples.
      The ASC crust corers were designed to recover, with each
deployment, a suite of discrete crust samples from specific locations
along a sampling traverse. The samples are of sufficient size to allow
chemical, structural, and thickness evaluation, as well as give an
indication of underlying substrate material. They were designed to be
deployed from a wide variety of oceanographic vessels using a standard
mechanical deep sea wire. The use of a mechanical rather than
conducting cable was preferred in order to keep the system development
and operation costs within realistic limits of available funding, and
make the system available for use on a wide variety of vessels.
      Two versions of the crust sampler have been built and tested at
sea. An initial prototype was built to test the concept of percussion
coring in the deep sea (Toth and Amerigian, 1986). This sampler
successfully recovered samples from the Hawaiian archipelago and led
the way for development of the Phase II system.
      The Phase II corer, shown in Figure 1, was designed to be used on
a routine basis for survey of manganese crust deposits. It has 30 core
guns plus support systems including:

(1)   electronic sequencing/firing system with bottom sensing trigger,

(2)   safety flasher/pressure shut-off switch,

(3)   35 mm camera system, and

(4)   acoustic telemetry pinger.

      The individually removable core guns are held in a gimballed frame
which is suspended within the steel vehicle cage. The electronics and
camera components are held in two steel frame pods bolted to the main
cage. The overall dimensions of the system are approximately 1.5 m h x
2.5 m w x 2.15 m d, with a weight of approximately 1200 kg in air. The
system is rated for operation to 5000 m depth.

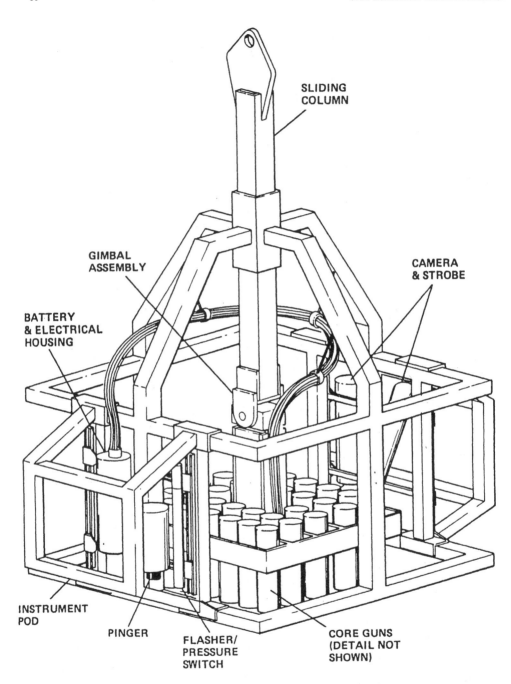

Figure 1.   Diagram of ASC Crust Sampler.

2.1. Coring Guns

The initial development of the crust sampler centred on the design of a
core gun incorporating a core barrel which would penetrate and retain
manganese crust material, and a mechanism for retaining the barrel and
retracting it into a protected enclosure to prevent loss of sample
during subsequent sampling with the remaining core guns. Initial tests
were made using components of a side wall coring system. Targets
consisted of large manganese nodules or pieces of coal embedded in
concrete, which simulated an expansive seafloor crust deposit. After
an extensive series of land-based and shallow water tests, a core gun
was developed which met the design criteria and could be mounted in
multiple units on the seafloor sampler.

A cut-away view of one of the core guns is shown in Figure 2. The
main components are a core gun body, from which the core barrel is
fired, core barrel, protective core gun tube, and a retraction
mechanism consisting of a retraction yoke, prestretched elastic cord,
tension rod and retraction trip lever.

The core body is a solid cylindrical piece of stainless steel with
a machined bore, firing chamber, ignitor seat and other features for
supporting the retraction mechanism. It is held in the core tube by a
single large diameter retaining pin. The core barrel is a 0.3 m long,
2.5 cm I.D. steel tube with a heavy round flange welded near its base.
This flange prevents the barrel from escaping from the core tube when
fired and is used to pull the barrel up into the tube by the retraction
mechanism. The core gun tube protects the retracted core barrel and
acts as a stand-off, allowing the barrel to reach maximum velocity
before it strikes the sample surface. It is 0.5 m long with a ring
welded inside the base of the tube to prevent the core barrel from
escaping and a square flange for mounting the assembled gun onto the
sampler vehicle.

The core retraction mechanism is activated by contact between the
core barrel flange and the base of the retraction yoke. This occurs
either upon firing of the barrel (if no sample surface is encountered
or the barrel penetrates to its full length) or upon raising of the
sampler and pull-out of the core barrel. When this occurs the yoke is
depressed slightly, the trip lever is released, and the yoke and core
barrel are pulled up into the core tube by the elastic cord.

The core barrel is fired using a pre-packaged pistol powder charge
of 20 to 40 grains which is fired by an electric ignitor. The ignitor
inserts into the top of the core body and is held in place by a screw-
in ignitor cap. All core gun components except the barrels are
fabricated from 316 or 17-4PH stainless steel. Core barrels are semi-
expendable, with a projected life of 10 to 15 deployments, and are
fabricated from 4130 chrome molybdenum steel.

2.2. Sampler Vehicle

The sampler vehicle consists of a main frame, sliding column with
gimbal for attachment to the core gun frame, and two instrument pods.
The main frame is a cage constructed of square tubing which provides
protection for the enclosed core guns and wiring harness and acts as a

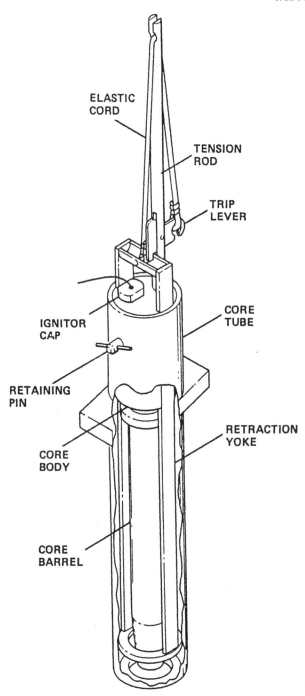

Figure 2.   Cut-away View of Assembled Core Gun.

platform for landing the sampler on the seafloor. The sliding column
allows the core guns to be brought into contact with the surface to be
sampled and the gimbal allows for conformance of the frame to
irregularities in the seafloor within the frame footprint. The
instrument pods protect the electronics and hold the camera, strobe,
and pinger away from the core gun frame so that they have unobstructed
visual and acoustic contact with the bottom.

The core gun frame is designed to hold and allow rapid loading and
unloading of the 30 core guns. The core guns slide into channels in
the frame where they are held in place. The wiring harness main cable
attaches near the centre of the frame and individual pigtails are
routed to each core gun inside closed channels. The frame has a square
central column which attaches to the gimbal on the central sliding
column.

2.3. Electronic Systems

The electronic systems are designed to

(1)   sequentially fire the core guns as the sampler is repeatedly set
      on the seafloor along its sampling traverse,

(2)   acoustically send data to the surface on the sampler altitude
      above bottom, mode of operation, and orientation,

(3)   take 35 mm photographs at each sampling site, and

(4)   lock out all sampler functions when ambient pressure is less than
      approximately 150 psi.

The system power supply and firing/sequencing electronics are
contained in a pressure housing which has an external switch for
resetting the electronics just prior to launch. The system ground is
routed through the safety flasher/pressure switch which holds the
circuit open, precluding firing of the core guns, and flashes the
strobe until an ambient pressure of approximately 150 psi (100 m water
depth) is reached.

The firing/sequencing electronics control all the major aspects of
the sampler operation, as follows: Upon positioning the sampler on the
bottom at a given sampling site, the system

(1)   waits for a continuous bottom contact signal for a 10 second
      period (to avoid false triggering),

(2)   outputs a firing pulse to the selected core gun,

(3)   sequences and fires up to 3 additional core guns at the given
      site, and

(4)   disables the bottom trigger for a pre-set time period (several
      minutes) to prevent accidental triggering during raising off the
      bottom.

In addition a continuous low voltage signal is output to the telemetry
pinger throughout the sampling sequence, causing it to alter its ping
rate. Prior to deployment selections are made for the number of guns
to be fired at each landing site (1 to 4), the time interval between
samples at each site (2 to 5 seconds), and the trigger lockout time
between sites (2 to 9 minutes), using dip switches on the electronics
board. Power for firing the core guns is supplied by a 30 volt
rechargeable lead-acid gel cell power pack.

Acoustic telemetry is sent to the surface by a 12 kHz pinger which
has been modified to output different ping rates according to the
status of the sampler. The normal ping rate is one second and in this
mode the pinger is used to monitor sampler height above bottom. During
the sampler firing sequence this ping rate is lengthened by a factor of
two. In addition if the pinger is tilted at an angle greater than 45
degrees from vertical the ping rate is lengthened by a factor of four.
Therefore when the sampler is operating successfully and is lowered
onto the seafloor the ping rate will shift from one to two seconds
while samples are being taken. A four second rate means the sampler is
lying on its side. An eight second rate means it is attempting to
sample while lying on its side. At each change in ping rate the ping
pattern is shifted by 1/2 second, which can easily be seen on a
shipboard 12 kHz line scan recorder.

3.   SEA TRIALS

The Phase II ASC Crust Corer was deployed for its first sea trials in
March 1987 from the research vessel Jean Charcot as part of a survey
cruise by the French government to locate potential crust mine sites in
the area of French Polynesia. Work at each site included a SeaBeam
bathymetric survey, side scan sonar survey, dredging, and coring with
the ASC sampler. The sites were relatively flat plateaus and dredge
recoveries prior to sampler deployment consisted of crust pieces and
nodules, indicating abundant but not necessarily continuous crust
pavement. The sampler was deployed at four sites with the most
successful deployment recovering 13 crust cores.

Overall, the sampler recovered crust cores in 34 percent of the
guns that were fired, returned damaged core barrels (indicating firing
into a hard rock surface) in 10% of the firings, and had no recoveries
in the remaining 56%. Colour 35 mm photographs were taken during each
deployment, however we have been unable to secure copies of these
photographs from the French to help evaluate sampler performance.
Considering the probable non-continuous nature of the crust and the
existence of ponded sediment at some locations, we consider these
performance figures quite good. The number of recoveries increased
steadily with each deployment (from 4 to 7 to 13 cores), until the
final deployment which experienced a failure in the sequencing
electronics.

Although all of the basic systems on the sampler operated well
during this first use at sea, two areas of operation were found to need
improvement:

(1)   greater core gun power needs to be available for deeper
       penetration into harder than average crust layers, and

(2)   improved electrical connectors are needed between the gun ignitors
       and wiring harness.

The crusts encountered on this cruise consisted of two distinct
layers; an outer layer generally 2 to 5 cm thick of higher grade
precipitated or "hydrogenous" material, and an inner layer up to 6 cm
thick of very hard, low grade material believed to be formed by
replacement of a carbonate substrate. The sampler easily penetrated
the hydrogenous layer but had difficulty penetrating more than 1 to 2
cm into the inner replacement layer. Attempts to use larger firing
charges for greater penetration resulted in damage to the rails holding
the core guns in place in the firing frame. Strengthening these rails
is a relatively simple matter which will allow greater charges to be
used.
       During the first deployment a problem was experienced with the
electrical continuity at the connections of the gun ignitors to the
wiring harness. This resulted in only 10 of the 30 core guns being
fired. This problem was overcome by effectively hard wiring these
connections for subsequent deployments and can be resolved for future
operations by replacing the current connectors with conventional single
pin underwater connectors.

4.   FUTURE DEVELOPMENT

ASC is moving ahead on two fronts with development of percussion coring
tools for site-specific sampling in the deep sea:

(1)   development of a single-sample untethered free fall device for
       recovery of crust and other consolidated substrates and

(2)   modification of the current sampler to recover samples of basaltic
       glass and interfacing the system with a conducting cable for
       remote operation from the surface.

4.1. Free-Fall Sampler

Free fall grab samplers have been used extensively by the marine mining
and academic communities for evaluation of manganese nodule deposits in
the central Pacific and other areas. Free fall samplers have distinct
advantages over tethered systems through increase of sampling
efficiency and ease of deployment and recovery. Typical deployment
consists of dropping a group of samplers at a given site in order to
obtain statistically relevant numbers of samples for assessment of
conditions at that site. Navigational control is tied to ship position
and sample position errors are therefore a function of sampler descent
rate, water depth and currents in the water column.
       Experience in evaluation of manganese nodule deposits has shown
that sample position control is very good with free fall samplers -

much greater than with most dredging operations. This positioning
control would be sufficient for evaluating most consolidated deposits,
with the possible exception of marine hydrothermal sulphides. Because
of the limited surface expression of most marine sulphides, we envision
that the free-fall sampler could provide samples from the general area
surrounding vent activity, recovering a suite of associated rock types
including altered basalt, oxide deposits, and sulphides if possible.
More precise determination of sampling positions could also be achieved
by outfitting a sampler with an acoustic relay transponder and
operating within a transponder array.

Funding for the first phase of development of this sampler has
just been received. The design goals include capability to operate in
water depths up to 6,000 m and slopes up to 45°, and two way travel
time and sampler loss rate equivalent to current free fall grab
samplers. Based on preliminary designs, the sampler will consist of
the following components:

(1) a single-shot core gun including bottom trigger and firing
     electronics,

(2) ballast and ballast release mechanism,

(3) glass or syntactic foam floats, and

(4) a core pullout mechanism.

The core gun mechanism will be similar to the individual guns used
in the ASC crust sampler. The core gun package will be made as light
as possible in order to minimize the sampler's float/ballast
requirements. The bottom trigger and firing electronics will consist
of a mechanical system to sense contact with the seafloor and a simple
electronic package which will fire the ignitor on command from the
trigger. A pressure-actuated safety lockout mechanism will be used to
prohibit firing of the core gun until it has reached a pre-determined
depth.

The ability to sample in rugged, locally high slope terrain will
be critical to the success of the sampler. Slopes encountered will
range from flat plateaus to vertical scarps on seamount flanks and mid-
ocean ridges. Maximum average slopes of seamount and ridge flanks,
measured with surface acoustic systems, range generally from 10 to 25
degrees, with local relief, from photographic and sidescan sonar data,
generally of a few centimetres to a few metres in scale. A design goal
of sampling slopes up to 45° has been selected, which we believe can be
reasonably achieved and is sufficient for most seafloor sites that will
be encountered.

Even more critical to the success of the sampler will be its
ability to return to the surface after sampling. This will be
accomplished by the use of a simple and reliable ballast release
mechanism and features to ensure successful core pull-out. Core pull-
out will be achieved by two mechanisms. Core barrels, which will
penetrate from less than 1 cm in basalt to 10 cm or greater in
manganese crust, will be designed to open a crater around the outside

of the barrel as they penetrate the target surface, allowing the barrel to pull free. In addition, systems will be investigated to assist in pulling the core barrel as the sampler begins its ascent to the surface. Such mechanisms will include:

(1) Release of a tethered flotation package, following firing of the core gun, which, as it reaches the end of its scope, would create a forceful pull on the core barrel.

(2) Forceful retraction of the core barrel into the core tube extension by a mechanism which would be driven by the inertia of the falling ballast weight(s).

The ballasting and flotation requirements of the sampler are a function of the submerged weight of the sampler components, the desired travel time in the water column, practical limitations of sampler size and weight with ballast for handling on deck, and ballast cost. The estimated submerged weight of the sampler, excluding flotation and ballast packages, will be approximately 20 kg. Based on our experience with Preussag and Benthos grab samplers, we would like to give the ballasted sampler a negative buoyancy in the range of 15 to 20 kg during ascent. This would allow round trip times of roughly 1 hour for each 2,500 m depth, depending heavily, of course, on the drag characteristics of the sampler.

4.2. Basalt Sampling

Funds are currently being sought by Lamont Doherty Geological Observatory to use the sampler for collecting precisely located samples of basaltic glass in conjunction with their geochemical studies on the East Pacific Rise. Modification of core barrels, attachment of a video camera, and operation from the surface through a conducting cable are planned, enabling highly precise, selective and well-documented sampling stations to be taken. Current plans call for the necessary sampler modifications to be made in early 1989 and testing to take place during Lamont's survey operation on the EPR that summer.

In the past, small but precisely located samples of basaltic material have been collected using a heavy, short barrel gravity core packed with grease to retain chips of rock. With current plasma spectrometry techniques, complete chemical analyses can be easily obtained on these types of samples. The percussion corer will be used to collect the same type of samples, but with the following advantages. The on-board video will allow greatly increased selectivity in sampling sites, and the sampler's multiple coring capability will considerably increase sampling efficiency.

Mechanical modifications for basalt sampling will centre on the design of new core barrels, which can be used with the existing core guns. Among the modifications that we will initially investigate will be the addition of expendable core cutters packed with a highly viscous substance such as grease or wax to retain the rock chips, and expendable cutters made up of a bundle of small diameter tubes which will collapse after impact and hold small pieces of sample. Land based

tests, similar to those made during development of the core gun, will
be conducted on these and other core barrel configurations in order to
finalize their design.

The sequencing and firing electronics will be split into a topside
deck unit, which will be used for manually firing the core guns, and a
sub-sea unit which will house the battery and sequencing electronics.
The video system will be mounted in the instrument pod in place of the
35 mm camera, where it will have an unobstructed view of the bottom for
surveying prior to landing.

5. CONCLUSIONS

Work conducted to date by Analytical Services Company has shown that
percussion coring is a viable technique for recovering site-specific
samples of consolidated marine deposits. The ASC crust corers have
successfully been used in surveys of manganese crust deposits in the
Hawaiian and Tuamotu archipelagos. Plans for future development work
include continued refinement of the current tethered crust corer,
interfacing of this system with a conducting cable for remote operation
from the surface and modification for collection of basaltic material,
and development of an untethered free fall percussion sampler.

6. REFERENCES

Lockwood, M.A. and Hill, G. (1986) Symposium Summary and
    Recommendations, Proceedings, The Exclusive Economic Zone
    Symposium Exploring the New Ocean Frontier, Washington DC,
    October 1985, pp.1-5.

Toth, J.R. and Amerigian, C.A. (1986) ´Development of an advanced
    sampling device for the investigation of marine ferromanganese
    crust deposits`, Proceedings, 18th Annual Offshore Technology
    Conference, pp.127-134.

Yuen, P.C., Corell, R., Craven, J., Takahashi, P. and Seymour, R
    (1986), unpublished draft, Symposium Summary and
    Recommendations, The Blue Revolution: Engineering Solutions for
    the Utilization of EEZ Resources, Honolulu, 28pp.

DEVELOPMENT OF UNMANNED SUBMERSIBLES FOR UNDERWATER OPERATIONS IN JAPAN

KENJI OKAMURA
Special Assistant to the Minister of State
for Science and Technology,
JAPAN.

Since the late 1970s, several kinds of unmanned submersibles for underwater observations had been developed in Japan and had been utilized successfully for the purpose of inspection in industries and also for scientific research. In the early 1980s, more advanced deep submergence unmanned vehicles have become required to support the underwater operations such as the construction, maintenance and repair works of submarine telecommunication cables in deeper water or scientific exploration of the deep seabottom or to assist the rescue operation of a deep manned submersible in an emergency case.

The KDD developed MARCAS-200 and MARCAS 200 CRAWLER in the early 80's for operations to the depth of 200 m. These submersibles have shown very successful results in many in-situ actual operations. The KDD has gained much knowledge and experience through these operations. However, the depth of the sea around Japan is quite deep and the fishing activities have become busy in the deeper sea more and more.

Submarine telecommunication cable system should have extremely high reliability and more than ten years MTBF (Mean Time Between Failure) is to be expected. However, in practice about 90% of cable failures were caused by external forces. About 60% of these failures were due to man-induced causes such as fishing gear and ship anchors, and almost 30% were due to natural induced causes such as turbidity current and tidal abrasion. In order to protect the submarine cable from fishing gear and anchors, cables installed in shallow water areas have generally been buried under the seabottom. The depth of buried cable under the seabottom was more than 1 m in some cases.

Subsequently the KDD developed the MARCAS-2500 in 1986, applying their high technological knowledge to perform the construction, maintenance and repair works for submarine cable installed in deep sea areas up to 2500 m.

The following functions are required for the unmanned vehicle:

1.   Visual inspection of the seabed and cables by TV cameras

2.   Determination of the seabottom profile and avoidance of obstacles

3.   Location and tracking of cables by magnetic sensors

D. A. Ardus and M. A. Champ (eds.), Ocean Resources, Vol. II, 69–76.
© 1990 *Kluwer Academic Publishers. Printed in the Netherlands.*

4.   Burial and reburial of cables by a jetting tool

5.   Measurement of sediment hardness and soil sampling

The maximum operating depth is 2500 m, the maximum operational sea
state is 5 and its tasks should include the measurement of burial depth
of cables, removal of obstacles such as fishing gear and anchors,
determination of faulty point of buried cables, and cutting the cables.
Considering the above requirements, the system configuration of MARCAS
2500 is illustrated in Fig. 1 and its principal particulars are shown
in Table 1.

TABLE 1. Principal particulars of MARCAS-2500

Operating Depth	2500 m
Dimensions	2.65 x 1.80 x 1.90 m
Weight	3.6 tf in air/-20 kgf in water
Structure	Titanium open frame
Pressure Housing	Titanium
Buoyant Material	Syntactic foam
Power Requirements	2250 v, 3 phase, 50 kw for electro-hydraulic unit. 1000 v, single-phase, 2 kw for electronics.
Propulsion	Six hydraulic thrusters two for forward-aft (12 ps each) two for lateral (5 ps each) two for vertical (7 ps each)
Instrumentation	LLL TV Camera on pan and tilt Stereo Colour TV Camera on pan & tilt B & W Cameras on pan & tilt B & W Cameras Still Camera on pan & tilt Four Floodlights Two AC Magnetic Sensors Acoustic TV on tilt Acoustic Locator, Altimeter, Gyrocompass & Magnetic Compass, Altitude sensor, Hydrophone, CTD sensor Responder for positioning, Pinger releaser, Hydraulic-pressure sensor, Oil temperature sensor
Attachments (optional)	Two manipulators Water Jetting Tool Sediment Tester Two DC magnetic sensors

   One of the most important matters is how to control the position
and the posture of the vehicle as required in various environmental
conditions without giving any extra load to the cable, considering the
hydrodynamics of the cable. The proper selection of size and weight of
the cable is essential. As the cable becomes heavier, the surface
tension will be increased, but the required propulsion power for

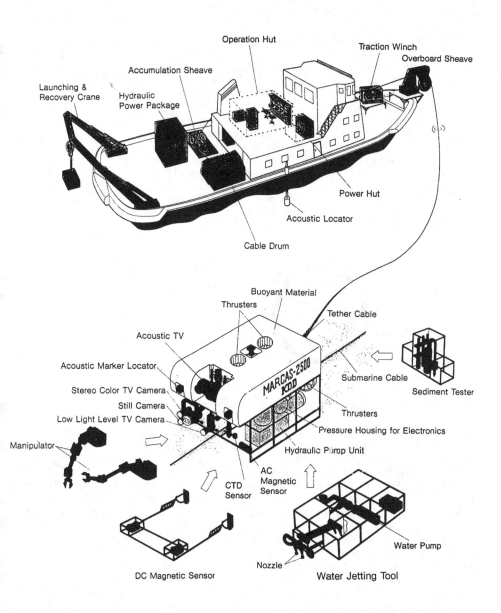

Figure 1. System configuration of MARCAS-2500.

vehicle will be decreased. The thinner the cable, the lower the
tension. The specification of the cable selected is shown in Table 2
and the cross section is in Fig. 2.

TABLE 2. Tether cable of MARCAS-2500

Outside Diameter	28 mm
Weight in Water	314 kgf/km
Cable Length	3500 m
Breaking Strength	Greater than 10 tf
Power Line	
3-phase	Resistance: 2.5 ohm/km
	Rated voltage: 2500 v
Single-phase	Resistance: 9 ohm/km
	Rated voltage: 1200 v
	Coaxial type
Optical Fibre	Three graded index fibres
Tension Member	Kevlar 49

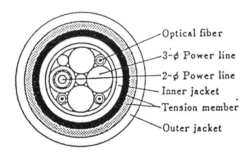

Figure 2. Cross section of tether cable of MARCAS-2500

Another important technology is the structure of the cable,
because the cable consists of many different materials such as, optical
fibre, power conductors, tension member, Nylon-loose tube, inner sheath
and outer sheath. Temperature coefficient for expansion, elastic
modulus and breaking strength are different for each material. The key
point is how to keep the optical fibre element from any deformation
while the cable is loaded and elongated with the load (nearly two tons)
which would be caused during the operations.

As for the automatic control of vehicle, sonar data such as
altitude, heading depth and angular acceleration are sent to the
surface computer that calculates proper voltages for controlling servo
valves of thrusters. An excellent performance of the autopilot was
confirmed during the sea trial, as shown in Fig. 3. The errors between
commands and measured values of depth, altitude and heading were within
±0.5 m, ±3 cm and ±1 degree.

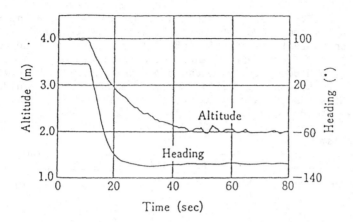

Figure 3.   Step response characteristics of auto-pilot (altitude and heading).

The fibre-optic signal transmission system is shown in the diagram in Fig. 4. The wavelength division multiplexing technique enables a high capacity of signal transmission with high quality. For the four-core optical rotary joint, the vehicle and surface transmission equipments are connected by three fibre-optic paths. Two of them are normally used and one is reserved for an emergency.

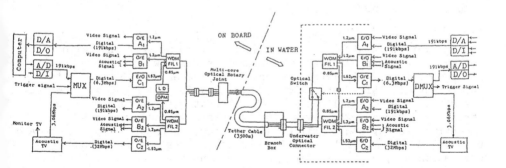

Figure 4.   Block diagram of signal transmission system.

The Japan Marine Science and Technology Centre developed its first
tethered vehicle JTV-1 and its series in 1980 and also the Hornet-500
with a cable using the optical fibre and optical wavelength-division-
multiplexing transmission system in 1984. Then JAMSTEC started to
construct Dolphin-3K, the largest and deepest tethered ROV (capable of
diving up to 3300 m deep) in 1984, and completed in 1987. This vehicle
was planned to be used for pre-site surveys, rescue of the manned
submersible Shinkai 2000, and scientific reconnaissance surveys.
Figure 5 shows the drawing of Dolphin-3K.

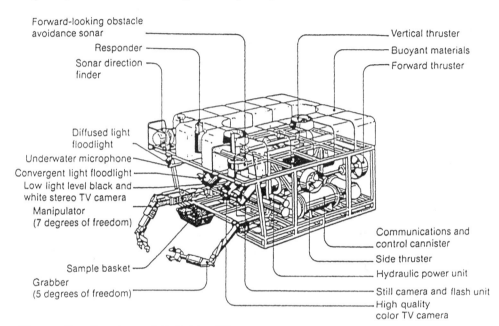

Figure 5. Drawing of Dolphin-3K.

The dimensions of Dolphin-3K are 2.85(L) x 1.94(W) x 1.96(H) m,
and the weight in air is 3700 kg and -10 kg in water. The payload is
150 kg. A 40 KW hydraulic system supplies the power to six thrusters
for propulsion, manipulator, grabber, cutters and pan & tilt units.
Instruments installed are colour TV camera, floodlights 5 x 500 W and
1 x 250 W, stereo still camera with 150 W strobe, current meter and
other instruments. For navigation, an obstacle avoidance sonar,
direction finding sonar, altimeter, gyrocompass angular rate sensor and
trim sensor are furnished. Shipboard components are a control/
navigation van, HV transformer and deck handling system. Tether cable
is of optical-electro-mechanical design, 30 mm in diameter and 5000 m
in length.
    The autopilot of Dolphin-3K is performed by PID control using some
linear filters for each gain. Autopilot functions include automatic up
and down control, automatic acoustic direction control, automatic
altitude/ depth control and automatic heading control.

Control and telemetry units are installed in a control van 6.4(L) x 2.4(W) x 2.4(H) m and are operated by two pilots. Dolphin-3K has demonstrated its excellent capability by picking up some unusual animals from very deep seabottoms.

A national project entitled "Advanced Robot Technology" under the R & D program of the Agency of Industrial Science and Technology (AIST), the MITI, is now underway. The target of this project is to establish those technologies for robots which could perform the inspection, maintenance and other complex tasks in environments that would not allow direct human intervention such as, maintenance jobs in nuclear power stations, undersea inspection of platform frames of offshore oil wells and disaster prevention in petrochemical plants. This project was started in 1983 and a prototype robot will be tested for overall evaluation in 1990.

This underwater system consists of a mobile robot, a relay base and support vessel. It will hover above a work site on the platform frame, settle on the frame, and then perform visual inspection, marine growth removal and cleaning and non-destructive inspection of welded portions. The movement of robot will be controlled automatically by pre-defined programs.

The size of the egg-shaped mobile robot will be about 3.2(L) x 2.7(W) x 2.7(H) m. The robot is required to keep its position and attitude in tidal currents within 0.15 m and 5°. The manipulator is required to handle instruments up to about 10 kgf in weight and also to keep its tip within a distance of less than 10 cm to perform various tasks. The mobile robot will be equipped with the following functional devices:

- three units of propelling mechanism for locomotion and hovering

- three legs with suction pad end effectors to fix on the frame

- dual arm manipulators with grippers

- sensors for navigation and positioning control

- computer systems for the control of locomotion and manipulator

- a stereo TV camera for operators on the support vessel

- an ultrasonic imaging device in turbid water

- acoustic communication network between robot and operators

- batteries for power supply

The utilization of ocean space in Japan is important and many underwater construction works are required. Recently the Port and Harbour Research Institute, MOT, has developed successfully in 1984 and 1987 the underwater walking robots named as "Aquarobot" for inspection purposes. ROVs suspended in the water are difficult to hold stationary in position and direction, and consequently are poor for the

measurement of the object and its position with high accuracy. This
robot is of a six-legged articulated "insect type". Each leg has three
articulations that are driven semi-directly by DC motors built into the
leg provided with sensors. All the motions are controlled by a
microcomputer and the robot can walk in any direction without changing
its quarters on the irregular rubble mound.

Recently, the Institute of Industrial Science, University of
Tokyo, has started the development of a free swimming vehicle for
exploration of deep seabottoms. The project is named "PTEROA".

PART II

Acoustic Sensors and Telemetry

HIGH-FREQUENCY COMMERCIAL SONARS: A SURVEY OF PERFORMANCE CAPABILITIES

CHESTER D. LOGGINS and WILLIAM J. ZEHNER
Sonatech Inc.
Santa Barbara
California
USA

1. INTRODUCTION

A number of "off-the-shelf" commercial sonars are available for a
variety of applications. When one has a requirement that calls for a
sonar, it is helpful to know which sonars in each category are
available, and what their performance capabilities are. Some
requirements cannot be met by existing sonars, and in this case it is
useful to know which companies have the potential, based on their
manufacture of similar equipment, to develop and manufacture a sonar
that will meet the requirements.
  The survey presented here is intended to provide information on
available high-frequency sonars: their performance capabilities; the
manufacturers of the sonars; and some general information regarding the
use of these sonars that involves fundamental, environmental, and
logistical factors that affect their performance. High frequency, as
used here, generally implies frequencies of 100 kHz or more, although
several of the sonars included in the survey have somewhat lower
frequencies. Only the fundamental features of the sonars as determined
by their acoustic and electronic design will be addressed in this
paper, and displays and post-processing systems will not be considered.
These equipments are very important to the overall performance of the
sonars, but at best they only preserve the inherent fidelity of the
information provided by the sonar. For the purpose of this paper the
sonars will be grouped into three categories: side-scan (or side-
looking) sonars; forward-looking sonars; and down-looking sonars.

2. PREDICTING THE ACOUSTIC PERFORMANCE OF ACTIVE SONARS

The parameters that determine the fundamental acoustic performance
capabilities of any active sonar are its operating frequency, its
horizontal and vertical beamwidths, its source level (SL), and the
bandwidth (B) and noise figure (NF) of its receiver. the sonar
equation for the received level, i.e. the acoustic level at the
hydrophone (dB relative to 1 micropascal), from a target of strength T
is

D. A. Ardus and M. A. Champ (eds.), Ocean Resources, Vol. II, 79–87.

$$LHT = SL - Losses + T$$

where SL is the acoustic source level (dB relative to 1 µPascal at
1 meter) produced by the sonar, and Losses is the transmission losses
term, which includes the spreading loss, 40LOG(R), and the absorption
loss, $2\alpha R$. The absorption coefficient α increases rapidly with sonar
operating frequency; it is the effect of this term that forces the
sonar engineer to use lower frequencies for longer range sonars.
    The source level achievable by a sonar is limited by several
factors. Ultimately, assuming that one can supply as much electrical
power to drive the transducer as needed, the power handling capability
of the transducer, cavitation, and the generation of nonlinear acoustic
waveforms limit the source level to less than about 230 dB.
    Whether or not a given sonar can detect a signal of a given level
depends upon the design of its receiver, since a noise that can mask
the signal is present, with an effective level of

$$N = -15 + 20LOG(f) + 10LOG(B) - DI - 10LOG(\eta) + NF$$

where f is the operating frequency in kHz, B is the bandwidth, DI is
the directivity index of the sonar (determined by the horizontal and
vertical beamwidths), NF is the receiver noise figure, and η is the
efficiency of the transducer material. This noise level defines the
minimum detectable level (MDL) of the sonar. A perfect receiver has a
noise figure of 0 dB, and good, real receivers have noise figures of a
few dB.
    In this paper a full discussion of acoustic performance prediction
cannot be presented, however the other three important acoustic signals
encountered in the operation of high-frequency sonars will at least be
mentioned. These are: volume reverberation; surface reverberation; and
bottom reverberation. Volume reverberation results from scattering of
the projected acoustic signal by particles within the isonified volume
of water, producing a return signal at the sonar hydrophone. In
shallow water, where the water depth is less than the maximum range of
the sonar, the water/air interface scatters the transmitted signal back
to the hydrophone, producing a surface reverberation signal. Surface
reverberation severely limits the performance of a side-scan sonar in
shallow water, particularly if the width, orientation, and sidelobe
levels of the vertical beam pattern are not properly controlled for
operation in shallow water. Bottom reverberation is the signal
backscattered to the hydrophone by the bottom. The predicted acoustic
performance of a nominal sonar, with design parameters typical of most
commercial side-scan sonars is presented in figure 1, and shows the
variation of all the important signal levels with range. The acoustic
performance of any active sonar can be accurately predicted when one
has the correct values for the source level and the minimum detectable
level of the sonar.

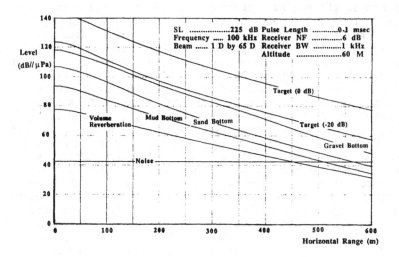

Figure 1. Predicted performance for Nominal Side-Scan Sonar.

3.   SIDE-SCAN SONARS

The side-scan sonar is perhaps the most common high-frequency sonar,
and is widely used for the location of small objects and for mapping
(imaging) the bottom.   There are three fundamental aspects of side-scan
performance important to the user of the sonar: its acoustic
performance, as discussed above; its imaging performance; and its area-
search rate.

As shown in figure 2 the side-scan sonar forms a beam that is
narrow in the along-track direction and wide in the vertical. The wide
vertical beam makes it possible for the side-scan sonar to search the
bottom from very short ranges out to the maximum range that the sonar
is capable, based on its source level and receiver design of achieving.

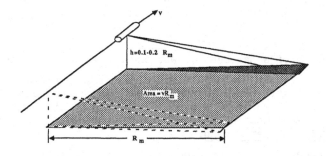

Figure 2. Side-Scan Sonar geometry.

        The fundamental imaging performance of the side-scan, and/or its
performance in locating objects on the bottom, is determined by its
horizontal beamwidth, and its resolution in the range direction. Range
resolution is determined by the pulse length and receiver bandwidth; in
a well-designed sonar the receiver bandwidth is the reciprocal of the
pulse length. This, unfortunately, is not always the case for
commercial side-scans, and results in receivers that have large noise
levels and, consequently, sonars with reduced range performance.
        The operating altitude of the high-frequency side-scan is
important to its imaging performance, affecting it in two ways. First,
the effective resolution in the horizontal range direction for a given
altitude varies with range, and is very poor near zero horizontal
range. This poor range resolution in the horizontal plane results in
an effective coverage gap (proportional to the operating altitude)
extending from a line on the bottom directly below the sonar, out to
the horizontal range where the effective range resolution (in the
horizontal plane) is approximately equal to the slant-range resolution.
A shorter pulse length can be used to decrease the width of the
effective gap in coverage some, but acoustic shadows become so short at
the shortest ranges that, even with the shorter pulse, the sonar
imaging performance is impaired and an effective gap still occurs.
        Second, shadows are important in interpreting the sonar image,
particularly in identifying targets of interest, and the operating
altitude of the side-scan affects shadow formation (specifically, the
length of the shadow); the rule-of-thumb for high-resolution side-scans
is to operate at an altitude of 10-20% of the maximum range.
        The final important measure of the performance of the side-scan
sonar is its area-search or area-coverage rate. As shown in figure 2,
a side-scan sonar moving at speed v, and having a maximum horizontal
range to each side of R, scans out an area equal to 2vR in unit time.
The speed of the sonar is limited by its horizontal beamwidth, and the
narrower the horizontal beamwidth, the lower the search speed of the
sonar.
        To provide a concise measure for use in comparing the performance
of various sonars in all three sonar categories, two new performance
parameters will be defined: the sonar information rate; and sonar
information density. The information rate provides a concise measure
of how fast a sonar searches, and is defined

        $$N_r = (n_b/\tau)\Gamma$$

where n_b is the number of sonar beams (two, one for each side, for all
commercial side-scans), τ is the pulse length, and Γ is the dynamic
range in bits. The dynamic range is the display dynamic range, and
will be assumed to have a value of 4 bits, since only about 16 grey
levels are discernible on most of the commercial sonar displays. The
units for N_r are kilobits per second.

        The second parameter is information density, and is just the sonar
information rate divided by the effective area-search rate of the
sonar. For the side-scan the information density is

$$N_\rho = n_b \Gamma / (2vR\tau)$$

where again Γ is the dynamic range in bits, v is sonar speed, R is the sonar maximum range, and τ is the sonar pulse length. The units of N_ρ for the side-scan sonar are <u>bits per square meter</u>: N_ρ gives the amount of information per unit area that is provided by the sonar.

The various high-frequency side-scan sonars are listed for comparison in figure 3. The operating frequency for many of these sonars is 100 kHz, the typical horizontal beamwidth is between 0.75 and 1.5 degrees, and the typical pulse length is 0.075 m (3 inches). Most of these 100 kHz sonars have range scales out to 500 or 600 m, and as figure 1 showed, these sonars should be capable of achieving such ranges. All of these sonars except the SLS-010 use a single-stave line array and perform no electronic beamforming, and except for the CMK-1 Shadowgraph and the SLS-010, are unfocused. Focusing becomes necessary when the beamwidth, as computed from $\theta = \lambda / L$, (λ = acoustic wavelength at the operating frequency, and L = array length), is less than about $0.5°$. Sonars for which resolutions better than $0.5°$ are claimed, and which are not focused, operate in the Fresnel region (near field) over their full swath. Their horizontal linear resolution, as defined by the linear extent of their horizontal beam pattern (measured at the -3 dB points), is essentially independent of range and has a value equal to the array length.

Sonar	Freq (kHz)	SL (dB)	Beam (deg) (hor by vert)	Pulse (meter)	Range Scale (meter)	Speed (knots)	ACR (sqnmi/hr)	Nr	Np	Vint	Manufacturer
EDO-601	100	224	1.3 by 57	0.075	to 500	15	2.5	20	1.5	75	EDO Western
EDO-606A	100	217	2.0 by 50	0.075	50 -400	15	5.1	60	4	83	EDO Western
EG & G 272-TD	105/500	228	1.2 by 50	0.075	25-600	10	2.4	80	20	85	EG &G
EG & G 990	59	220	1.2 by 40	0.075	100-500	10	2.4	40	10	--	EG & G
Star Scan	100	224	1.5 by 65	0.075	25-600	15	4.4	60	6.5	80	Electrospace Systems
Multi-Scan 1500	100	229	1.0 by 55	0.075	----	--	--	--	--	79	Ferranti Ocean Res Eq.
Sea Marc I	27/30	228	1.7 by 50	0.2-3.2	500, 1000,2500	7	1.7	40	3	--	I.S.T., Inc
Sea Marc CL	150	--	1.5 by --	R/1024	25-500	7	1.4	40	8	--	I.S.T., Inc.
Klein 422S-101AF	100	228	1.0 by 40	0.075	25-600	12.7	2.1	40	5	--	Klein
Klein 422S-101BF	100	228	1.5 by 40	0.075	25-600	16	3.3	40	5	--	Klein
Klein 422S-101EF	500	216	0.2 by 40	0.015	25-600	2.5	0.14	200	350	--	Klein
Klein 422S-101GF	50	228	1.5 by 40	0.15	25-600	16	3.3	20	2.5	--	Klein
Klein 422XZ-101AF	100	228	1.0 by 40	0.075	25-600	12.7	2.1	40	5	--	Klein
Klein MK 24 Mod0	100	228	0.75 by 40	0.075	25-600	9.5	1.5	40	7	80	Klein
SLS-010	600	220	3 (0.2m by 70d)	0.2	37.5,75,100	12	0.97	192	200	88	Sonatech, Inc.
CMK-1 Shadowgraph	1200	--	0.1 by 90	0.038	80 feet	3	0.08	40	518	58	Westinghouse
Unknown	170/190	---	0.5 by 90	0.3	600	2	1.2	10	4	--	Thomson CSF
Waverley 3000	100	227	1.5 by 50	0.075	75-600	10	2.7	40	10	--	Waverley Elect., Ltd
Wesmar 500SS	105	---	1.5 by 35	0.075	30-480	2	1	60	59	79	WESMAR

Figure 3. Side-Scan Sonars.

The CMK-1 Shadowgraph sonar has a unique design that provides focusing using an array that is an arc of a 15-foot radius circle. So long as the sonar is operated at an altitude of 15 feet, simple geometry ensures that the beam is perfectly focused at every range, from zero to infinity, on a horizontal plane 15 feet below the sonar. The Shadowgraph produces very high quality side-scan images, but has a short maximum range because of its very high operating frequency (1.2 MHz). The SLS-010 sonar built by Sonatech is also focused, although in this case the focusing is applied in a digital beamformer. The SLS-010 is completely digitally beamformed, and features all-range electronic focusing, range-varied aperture length beamforming to maintain constant linear along-track resolution, and multiple simultaneous beams for high area-search rate performance.

The information presented in figure 3 is based on manufacturers' claims and advertising. The MDL's of the sonars are not usually specified by the manufacturers, and the acoustic performance provided by the sonars cannot, therefore, be accurately estimated. The area-search rate values were computed under the constraint that the tow speed must not exceed a value that produces more than a 50% loss in coverage caused by "holidays" (unsearched areas that result from the convergence of the horizontal beam at short ranges, not to be confused with the gap discussed previously). Figure 3 may not include all of the most recent models produced by the manufacturers.

4. FORWARD-LOOK SONARS

Forward-look sonars are used for obstacle detection and avoidance, fish finding, and/or for the surveillance of an area from a stationary platform or location. These sonars use either a single beam that is mechanically scanned to cover a desired field-of-view (FOV), or multiple, pre-formed beams that scan the FOV in a single pulse period; these are called SWAP (scan within a pulse) sonars. There is more variety in the design and configuration of forward-look sonars than for sonars in either of the other two categories; there are mechanically scanned pulsed sonars; mechanically-scanned CTFM (continuous transmission frequency modulation) sonars; SWAP sonars that cover a one-dimensional FOV; and SWAP sonars that cover a two-dimensional FOV.

The details involved in the design and operation of these various types of sonars cannot be discussed here, but some general comments and comparisons can be made. The mechanically scanned sonars are clearly more suitable for operation from a stationary, or very slowly moving, platform since a large number of pulses is required to scan the FOV, and movement of the sonar between pings distorts the sonar image. The CTFM sonars provide continuous, simultaneous data from all sonar range cells, but single beam CTFM sonars still suffer from image distortion when operated on a moving platform. The sonar image produced by a SWAP sonar is undistorted because it is produced by a single ping, and the SWAP sonar is the sonar best suited for operation on a moving platform.

Some of the important parameters of forward-look sonars are: the number of pre-formed beams; the beam width (vertical and horizontal); the scan rate; the FOV; the pulse width, or, in the case of CTFM, the number of range filters or bins; the design type (CTFM, SWAP, mechanically-scanned); and the important acoustic parameters (source level and MDL). The acoustic performance of the forward-look sonar is predicted in the same manner described already. The imaging performance of these sonars is determined by the number and width of the pre-formed beams, the scan rate, and the pulse width (or range bin width).

The information rate for non-CTFM forward-look is calculated using the expression as before, where n_b is the number of pre-formed beams. For CTFM sonars, the information rate is given by

$$N_r = (n_f)(W_f)(n_b)\Gamma$$

where n_f is the number of filters (each which corresponds to a range bin) in the frequency analyzer, and W_f is the filter bandwidth. The information density for a pulsed, forward-look sonar is given by

$$N_\rho = \Gamma/(\Delta r \theta_h)$$

where Δr is the range resolution in meters, and θ_h is the horizontal beamwidth. For the CTFM sonar, N_ρ is given by

$$N_\rho = \Gamma W_t/(R_{max} W_f \theta_h)$$

where W_t is the total bandwidth of the frequency analyzer. The units of N_ρ for forward-look sonars are bits per meter-radian.

The various forward-look sonars are compared in figure 4. The simplest forward-look sonars, the mechanically-scanned, pulsed sonars, are listed first. These sonars, which have the lowest scan rates of all the forward-lookers, are used for a variety of applications, from fish finding to bottom profiling. Next in level of complexity are the mechanically-scanned CTFM sonars. The multi-beam, pulse sonars (SWAP sonars) are listed next and have the highest scan rates.

Three of the forward-look sonars listed are unique in design and/or features and merit special discussion. The TOAS (Terrain and Obstacle Avoidance) sonar is the only forward-looker that provides discrimination in both the horizontal and the vertical planes. TOAS forms a matrix of 15 beams (each 11° x 11°) that cover a 55° wide by 33° high FOV. TOAS was developed to provide underwater vehicles with an obstacle and terrain avoidance capability, and this application does not require high-resolution beams, hence the relatively broad beams.

The second sonar with a unique design is the Plessey Mirror sonar. This sonar uses a specially-shaped mirror to focus the acoustic signals on an array of transducer elements to form a fan of high-resolution beams. No beamforming electronics are required since the mirror is the beamformer. The mirror is independent of frequency (over a wide range of frequencies) and the sonar has two different operating frequencies.

The third unique forward-looker is the EAARS 90 F.F. sonar made by Ulvertech. The unit uses a monopulse technique to resolve the directions of targets within its FOV. In a monopulse sonar, the angular resolution is dependent upon the signal-to-noise ratio (SNR), and when the SNR is large, the sonar can resolve target direction extremely accurately. Multiple targets, in a single range cell, affect

the performance of the monopulse algorithm and can result in an
incorrect angle.

Sonar	Type	Freq(kHZ)	FOV	Proj Bpat	Nb	Hydro Bpat	RR	SL	Nr	Np	Applic.	Scn Rt	Vin	Manufacturer
FH-106	MS,P	60/150	360 by --	-- by --	1	-- by --	--	--	--	--	FF	--	--	Furuno
MESOTECH 971	MS,P	675	360 by 30	1.7 by 30	1	1.7 by 30	0.075	--	40	1800	BP,OA	1.7	--	Mesotech
UDI AS 360	MS,P	500	270 by 27	1.3 by 27	1	2.7 by 27	0.075	--	40	2050	OL & S	7.2	--	UDI Group
WESMAR 100RV	MS,P	266	360 by 8	8 by 8	1	8 by 8	.15-1.9	224	--	--	NA, RV	--	--	WESMAR
WESMAR QDS265	MS,P	160	360 by 6.5	6.5 by 6.5	1	6.5 by 6.5	---	229	--	--	HS,D	--	--	WESMAR
EDO 4059 OAS-1	MS,P	100	180 by 50	2 by 50	1	2 by 50	.07-.35	218	40	1530	OA & S	5.8	80	EDO Western
AMETEK 250A/255	MS,C	107-122	360 by 15	44 by 15	1	3 by 15	0.12	185	7.5	1.5	OA & S	30	78	Ametek-Straza
AMETEK 500A/550	MS,C	87-92	360 by 15	60 by 17	1	2.5 by 15	9.1	190	2.5	2.5	OA & S	25	68	Ametek-Straza
AMETEK DHS-2	MS,C	116-95	10 by 10	18 by 18	1	10 by 10	--	180	--	--	Diver	--	--	Ametek-Straza
AMETEK 300	MB,P	200	120 by 18	120 by 18	40	4 by 18	0.38	200	80	50	OA	450	74	Ametek-Straza
C-TECH CDS-40A	MB,P	36	360 by 9	360 by 9	36	12 by 9	0.75	210	108	23	FF	540	--	C-Tech Ltd
C-TECH CMS 5000	MB,P	30	360 by 9	360 by 9	36	12 by 9	1.35	214	80	16	FF	540	80	C-Tech Ltd
C-TECH LSS-68	MB,P	75	200 by 9	200 by 9	20	12 by 9	0.75	217	20	7.6	FF	300	78	C-Tech Ltd
CH-12	MB,P	155	30 by 4.5	30 by 4.5	6	6 by 4.5	0.53	--	--	--	FF	21	--	Furuno
CS-50 MK2	MB,P	55	360 by --	-- by --	--	-- by --	--	--	--	--	FF	--	--	Furuno
KRUPP 950	MB,P	19.5	90 by 14	9 by 14	12	9 by 14	1.5	223	12	11	FF	135	--	Krupp
MARCONI 360	MB,P	83/250	360 by 15	360 by 15	110	3.3 by 15	0.1	--	3385	690	SURV	900	80	Marconi
Marconi Hydrosearch	MB,P	180	60 by --	60 by --	120	0.5 by --	0.075	--	--	--	OL&HS	--	83	Marconi
T.O.A.S.	MB,P	200	55 by 33	60 by 55	15	11 by 11	0.75	196	220	144	OA&TA	330	86	Sonatech
Plessey Mirror	MB,P		SPECS	AVAILABLE		ON REQUEST								Plessey
EAARS 90 F.F.	Mono	200	90 by 4	90 by 4	1	90 by 4	0.06	--	50	42	EA & R	210	--	Ulvertech

FF-fish finding   NA-navigation aid   EA-electronic avoidance
BP-bottom profiling   HS-harbor surveillance   R-reconaissance
OA-obstacle Avoidance   SURV-surveillance   OL-object location
S-Search   TA-terrain avoidance   SD-swimmer detection

Figure 4. Forward-Look Sonars.

5. DOWN-LOOKING SONARS

Down-looking sonars comprise the smallest category of the sonars
surveyed. The down-looking sonar is generally used for bottom contour
mapping, depth sounding, and fish finding. These sonars use a beam
aimed directly below, or a fan of beams centered on a direction
directly below the vehicle. In the single-beam systems, the beam is
similar to the side-scan sonar beam in that it is relatively wide in
the vertical, and narrow in the horizontal (the direction of vehicle
motion). The multi-beam down-lookers use a fan of beams in the
vertical plane to derive bathymetric or bottom contour information over
a fairly wide swath to either side of the sonar. These sonars are
essentially multi-directional depth sounders, determining depth at all
horizontal ranges within the swath with a single ping.

    The down-looking sonars are listed in figure 5, and they cover a
wide range of frequencies and performance, from 1 MHz down to 12 kHz,
and operating ranges from 75 m out to 1100 m. Most of the down-lookers
are mounted on surface platforms, and only the Benigraph and the SIMRAD
sonars use a towed vehicle. The multi-beam down-lookers are similar in
principle to the multi-beam, pulsed forward-looking sonars, while the
single-beam down-lookers are very similar to a one-sided, side-scan
sonar.

Sonar	Freq (kHz)	SL (dB)	FOV(deg) (hor by vert)	PulseLength (meter)	Nb	Hydro. BPattern (h deg by v deg)	Vint	Manufacturer
Benigraph	1000	--	0.5 by 100	0.05	87	0.5 by 0.57	--	Bentech
Hydro Chart	36	--	5 by 105	0.75	11	5 by 30	82	General Instrument
Sea Beam	12	--	2.7 by 44	--	16	2.7 by 20	82	General Instrument
SIMRAD EM100	95	--	3 by 80	0.5	32	3 by 2.5	85	Simrad Subsea
WESMAR VS3000	200	--	-- by --	0.075	1	-- by --	--	Wesmar

Figure 5.   Down-Looking Sonars.

6.   SUMMARY

A number of high-frequency, commercial sonar systems have been surveyed
and listed according to three fundamental categories for comparison of
performance features.   This survey provides the user, or potential
user, of such sonar systems with information useful for selecting a
sonar for a given application, or in the event none of the sonars
available meet the user's requirements, identifies manufacturers who
have the potential to design and build a sonar that will meet the
user's needs.

U.U.V. ACOUSTIC SENSORS

GARY L. BANE
Rockwell International Corporation
Autonetics Marine Systems Division
3370 Miraloma Avenue, P.O. Box 4921
Anaheim, California 92808-4921, USA

1. INTRODUCTION

Sensors for Unmanned Underwater Vehicles (UUVs), in support of various missions, will generally consist of acoustical sensors and processors since the medium for signal/data transmission and reception is water. As such, this paper will necessarily concentrate on such sensors/processors with state-of-the-art technology and advancement projections to the Year 2000.

The offboard host system, called an Unmanned Underwater Vehicle (UUV) can be free-swimming with varying degrees of autonomy (Autonomous Underwater Vehicle - AUV). It can also be tethered to a mother platform by relatively short electrical/power umbilicals (Remotely Operated Vehicle - ROV); or, it can be "connected" by a longer data link only, such as fiber optic links, RF transmissions, or acoustic signals (Untethered ROV - UROV). The first type, the AUV, must have significant onboard artificial intelligence while the latter two do not.

For the purpose of this paper, a single mission will be established to best examine the sensor issue for UUVs. This mission is one of ocean bottom surveys.

Clearly, no single sensor exists or is anticipated that will do the total job. A synergistic application of multiple sensors and an understanding of the environment will always be required, as depicted in figure 1.

In summary, many acoustic sensor components and devices for use in the deep ocean have been developed. A substantial technology base exists at the component level which is common to all such systems. State-of-the-art will be defined for such systems as well as projections as to what is achievable in a dozen years (to the Year 2000) with engineering development and what will be achievable with research and development.

D. A. Ardus and M. A. Champ (eds.), Ocean Resources, Vol. II, 89–104.
© 1990 *Kluwer Academic Publishers. Printed in the Netherlands.*

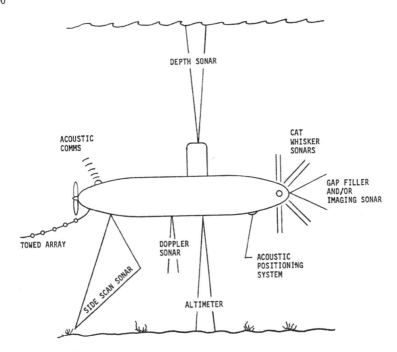

Figure 1.   AUV acoustic sensors.

2.   THE ACOUSTIC SENSORS

The acoustic sensors required for most UUV missions and especially for
the ocean bottom survey mission can be grouped into two areas for
convenience:

1.   Navigation

2.   Sonar

2.1 Navigation

Typical navigation system performance is usually stated in terms of
either relative accuracy or absolute (geodetic) accuracy. Under proper
conditions, very high relative accuracies can be achieved. This is
accomplished by operating in a coordinate grid that is limited to the
combined navigation systems, its platform, and other navigational aids
such as acoustic transponder fields. Without a geodetic reference, it
is impossible to transfer accurate coordinate data to another grid
system. Geodetic accuracy is referenced to a world earth model and
presents no hand-over problems within a local datum.
     Inertial navigation systems, which have advantages over other
technologies (such as detectability in acoustic transmissions) are
self-contained, unradiating, and unjammable. Ring laser gyros in

navigation systems are expected to be available by the early 1990s, furnishing UUVs with accuracies of 0.05 nm/hour within 1 ft^3. Doppler sonar is a navigation aid operated in conjunction with acoustic or inertial systems. It is used most often to check the integrated inertial system results. When operated alone for local relative navigation, its accuracy is 0.2% of the distance travelled, plus a 60 ft/hour drift. Directionality, low power, and short ranges prevent its detectability.

The current navigation options for UUVs are pictorially shown in figure 2. The current state-of-the-art of such approaches is shown in Table 1.

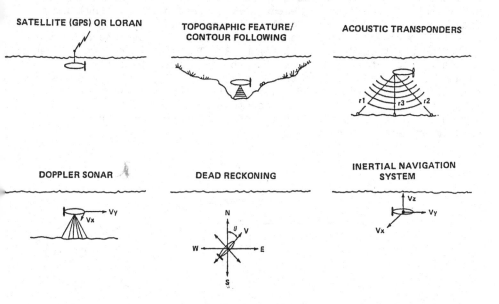

Figure 2.   Autonomous navigation methods.

TABLE 1.   Navigation systems.

ACOUSTIC	Requires transponders, highly accurate
DOPPLER/COMPASS	>0.2% of distance travelled
INERTIAL	0.00014 NMI/HR Drift 23 ft^3 0.018 NMI/HR - 2 ft^3 1 NMI/HR - 1 ft^3
LORAN C	1000 ft absolute - TRANSIT
SATELLITE	700 ft absolute - TRANSIT 50 ft - GPS

        A future concept conceived by Rockwell, called the ALAG
(Autonomous Local Area Guidance) system is briefly discussed to
highlight the integration of various sensors to resolve the guidance
and control requirements of the 21st century UUV.
        The control and guidance of an autonomous vehicle has two major
functions, the first is the determination of its current position and
from that point establishing a plan to reach its desired destination.
The second function is to sense the environment in the vicinity of the
vehicle and to guide the vehicle to meet a "local" objective. The
local objective may be to maintain a constant depth of attitude,
station keeping with another vehicle or simply to avoid colliding with
objects. The navigation and guidance of a UUV can be summarized as a
global problem (determine present position and path planning) and a
local problem (sensing the local environment and providing "safe"
steering commands). These tasks can be related to the operation of a
submarine which is depicted in figure 3. The global problem is that of
the navigation officer and captain, who establish the track plan. The
local problem is handled by the Officer of the Day (OOD), who provides
commands to the helmsman and diving officer for control of the
submarine. The concept is an Autonomous Local Area Guidance (ALAG)
system which performs the function of the OOD for an Autonomous
Underwater Vehicle (AUV).

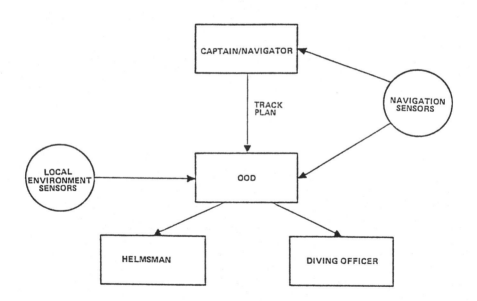

Figure 3. Submarine guidance and control.

The types of sensors which would be required for the ALAG system are shown in figure 4. The obstacle avoidance sensor would be a multibeam system to allow intelligent turn decisions to be made when obstacles are encountered. The required range and area of coverage for this sensor is dependent upon the speed and maneuverability characteristics of the vehicle and the surface (wind) and bottom (sand, mud, etc.) conditions in the area. The purpose of the high resolution sensor shown in figure 4 will be to detect objects, classify them and ultimately to provide the ability to grasp or attach something to the object. The bottom and ice sensors could be conventional sensors now in use in underwater vehicles; however, some improvements may be desirable for resolution, increased range and sensitivity (mud bottoms).

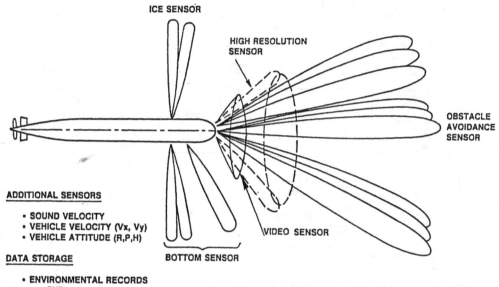

Figure 4.   ALAG sensors.

A block diagram of the navigation and guidance function for an AUV is shown in figure 5. with the ALAG system shown enclosed in a dashed line. A comparison of figures 5 and 3 indicates the equivalence of the manned versus AUV navigation and guidance. The vehicle command/control function (CAPTAIN/NAVIGATOR) will establish the guidance objectives for the ALAG system. The normal operating mode while transiting will be to travel a straight line from point A to B. The distance from A to B can be varied to cause the vehicle to travel specific trajectories such as a great circle path. Some of the other available guidance objectives are maintenance of a specific heading (either vehicle pointing or velocity vector), terrain following, contour following, station keeping or target closure. The ALAG system (OOD) would utilize its local environment sensors to provide commands to the vehicle control system (helmsman and diving officer). Accomplishing the ALAG objectives may require access to records of the expected environment, and this data base must be updated during operations to reflect the observed environment.

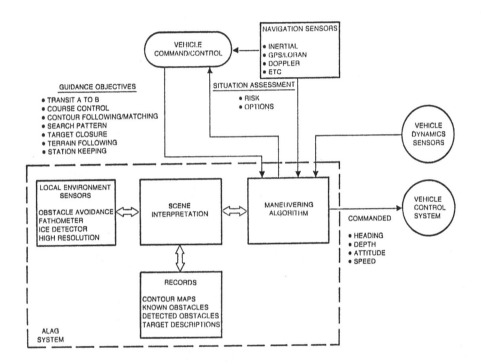

Figure 5.   ALAG system block diagram.

The ALAG system must also provide some inputs to the command/control function to keep this function informed of the current situation.

3.   SONAR

Advanced state-of-the-art sonars for UUVs are under development and in varying stages of hardware completion and testing. Three particular and unique types, designed specifically for the ocean bottom survey UUV mission, are:

(a)   a forward looking, obstacle and terrain avoidance sonar;

(b) an ahead and downward looking sonar to "fill-the-gap" of the UUV
     mounted side-looking sonar; and

(c) a digital side-scan sonar.

3.1 Forward-Looking Sonar

The TOAS (Terrain and Obstacle Avoidance Sonar) is a small, lightweight
forward-looking sonar designed specifically to provide unmanned
underwater vehicles with a capability for terrain and obstacle
avoidance. TOAS forms a matrix of fifteen preformed beams. Each beam
is 11° by 11° wide, and has sidelobes 30 dB below the mainlobe. This
system of beams defines the vertical and horizontal positions of
obstacles in the path of the vehicle, the basic information needed for
taking correct evasive action. TOAS operates at a frequency of
200 kHz.

     TOAS consists of four curved arrays, one projector, and three
identical receive assemblies arranged to cover a 55° horizontal x 33°
vertical forward matrix consisting of 5 horizontal x 3 vertical 11°
cells. An integral acoustic processor accepts functional commands from
an external detection processor to select power, TVG, variable gain,
pulse width and pulse repetition rate, and hand back detected envelope
data. It can interface high information rate acoustic signals to the
artificial intelligence systems being developed by Rockwell
International for AUVs. The lightweight, low-powered sonar uses
preformed, spatially oriented beams to acquire forward-looking obstacle
and terrain returns. These are processed by the vehicle and used to
provide steering information and navigation decision making. Embedded
processors control the many setable parameters of the system to tailor
acoustic functions to varying vehicle environments.

     The hydrophone and analog processor are detailed in figure 6. On
command from the digital processor, a 200 kHz pulse is applied to the
power amplifier input, amplified, and sent to the acoustic projector to
ensonify a 50° x 60° sector forward of the vehicle. Both the ping
width and source level are selectable. Target returns are received by
15 beams, each of which is 11° x 11°. The received signals pass
through two amplifiers; a variable gain amplifier, and a time-varying
gain (TVG) amplifier. After amplification, the signal is bandpass
filtered, envelope detected, and sent (in analog form) to the post
processor. The variable gain amplifier settings and TVG curves are
both selectable and controlled by the post processor. The filtered and
detected analog signals sent to the digital processor are multiplexed
and digitized at a 6 kHz rate. The largest amplitude of each six
sequential data points is selected and compared to a set of time
varying thresholds. These data points (target candidates) are tracked
over a number of ping cycles and classified as targets or non-targets.
Non-targets are dropped from the tracking list, and targets are
characterized by range and direction (receive beam number) and sent to
the host vehicle to be used as inputs to manoeuvring algorithms.

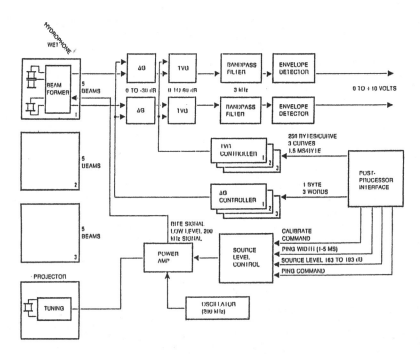

Figure 6. TOAS hydrophone and analog processor functional block
diagram.

3.2 The GAP Filler Sonar

**THIS SECTION CONTAINS PROPRIETARY INFORMATION TO SONATECH INC., AND
SUCH INFORMATION MAY NOT BE USED FOR MANUFACTURING PURPOSES NOR
DISSEMINATED BEYOND THE CONFERENCE ON INTERNATIONAL OCEAN TECHNOLOGY
CONGRESS, EEZ RESOURCES: TECHNOLOGY ASSESSMENT.**

     Primary target classification clues are provided by the acoustic
shadow cast on the bottom by targets of interest. Targets located
immediately beneath the UUV, however, are not easily observed because:

(a) the high levels of reverberation from the bottom at near-normal
     incidence mask the target highlights, and

(b) the geometry prevents shadow formation.

The strip of bottom directly beneath the UUV can be searched using an
ahead-looking sonar mounted in the nose and operated at low vehicle
altitudes to produce low grazing angles and thereby maximize shadow
formation.
     The design requirements for this sonar depend heavily on the
definition of what constitutes a shadow and how the sonar suite is
employed. If one assumes that:

(a) the transmitted pulse length is matched to the size of the target
     of interest,

(b) the sonar is flown at an altitude of 20% of maximum range scale
     (R_{max}), and

(c) the target shadow length must be at least 3 range samples in
     length,

then it can be shown that the side-scan sonar output is only useful
from 60% of R_{max} to R_{max}. To fill the resulting gap requires the
forward-look sonar to operate at a range of at least 0.6 R_{max}. To
provide resolution equivalent to that of the side-scan sonars, this
sonar must have 400 beams for the long-range side-scan sonar case and
1440 beams for the short-range sonar case. Forward-look sonars having
such performance are clearly not practical.

The sonar search geometry under the assumptions just stated is
illustrated in figure 7(a). This illustration shows the area covered
by the forward-looker and the side-scan sonar.

The complexity of the required forward-look sonar is greatly
reduced when the assumptions regarding shadow formation are relaxed and
the operating altitude of the sonar is reduced. For example, if we
require a minimum shadow length of only one range sample, and the
vehicle is operated at 10% of R_{max} instead of 20%, the maximum range
and number of beams required for the forward look become more
reasonable: 67 beams for the long-range sonar case and 240 beams for
the short-range sonar case. Even under these relaxed assumptions, the
required forward-look sonar is impractical. The new search geometry is
illustrated in figure 7(b).

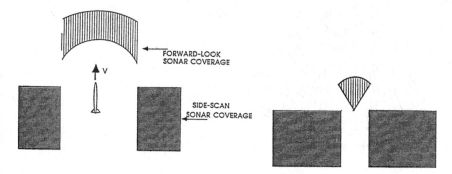

(a) PLAN VIEW OF SEARCH GEOMETRY FOR THREE CELL LONG   (b) PLAN VIEW OF SEARCH GEOMETRY FOR ONE CELL LONG
    SHADOW AND ALTITUDE 20% OF Rmax   SHADOW AND ALTITUDE OF 10 % OF Rmax

Figure 7. Search geometry for Gap-Filler Sonar.

Rather than attempt to design a forward-look sonar having the
side-looking sonar resolution performance, it is felt that the multiple
looks at targets obtained within the sector covered by the forward-look
sonar will provide good detection performance in the gap between the
port and starboard side-scan sonars, providing effective gap-filling.

A forward-looking sonar for filling the gap left by the port and
starboard side-scan sonars will be a 64-channel line array, with an
FFT-beamformer. This sonar will operate at the nominal frequency of
150 kHz and will cover a sector 60° wide centered on the UUV heading
direction. This sonar will rely on an individual target being seen
many times as the vehicle moves forward to provide a high probability
of detection.

The FFT-beamformed line array produces 30 beams, each 2° wide,
covering a 60° sector centered on the heading direction of the sonar
platform. A separate projector transducer insonifies the sector to be
searched. This projector is shaded both horizontally and vertically to
produce a beam pattern with very low sidelobes in both planes. The
beam patterns of the FFT-beamformed forward-look sonar are illustrated
in figure 8.

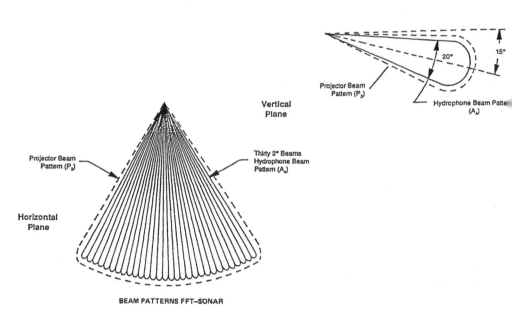

Figure 8.   Forward-Looking Sonar beam patterns.

The acoustic performance of the FFT-beamformed forward-look sonar was computed. The results predict that a -20 dB target can be detected out to ranges greater than 300 m. The probability of detection for a single-look for a target at this range is only around 0.4, but this will be sufficient because each target will be seen many times, around 80 times at 4 knots, producing a very high cumulative probability of detection. The forward-look sonar can perform well using a poorer resolution than a side-scan sonar because it provides many consecutive looks at each object within its field-of-view.

3.3 Side Scan Sonar

Digital Side-Scan Sonar

THIS SECTION CONTAINS PROPRIETARY INFORMATION TO SONATECH INC., AND SUCH INFORMATION MAY NOT BE USED FOR MANUFACTURING PURPOSES NOR DISSEMINATED BEYOND THE CONFERENCE ON INTERNATIONAL OCEAN TECHNOLOGY CONGRESS, EEZ RESOURCES: TECHNOLOGY ASSESSMENT.

The speed and flexibility of off-the-shelf Digital Signal Processing (DSP) hardware make it possible to build a new generation of high-performance side-scan sonars. They will provide greatly enhanced performance capabilities and versatility. These sonars will form multiple high-resolution beams employing range-varied aperture length, range-varied focusing and range-varied steering.

The inherent versatility provided by the digital design approach describe below, coupled with power of available digital hardware, also makes it possible to operate at several different frequencies, ranges, and resolutions - providing both detection and enhanced classification capability with a single sonar.

3.4 Beamforming Approach

Beamforming for high-resolution side-scan sonars is performed using phase-shift operations and true time-delay techniques are not required. A rare exception to this rule is the synthetic aperture sonar which, because it forms enormously long (synthetic) arrays, does require the time-delay beamforming.

The basic phase-shift operation can be implemented in hardware in several ways. For a digital beamforming approach, however, the most straightforward implementation is just the digital complex multiply.

The phase correction values represent constant relative interchannel phase errors measured in an acoustic calibration procedure illustrated in figure 9. The basic principle involved in this calibration is that a source (or a strong "point" target) at a range R from the array produces a spherical wavefront, with a radius-of-curvature of R, at the face of the array. The range must be large enough so that all array elements "see" the source within the 3 dB limits of their mainlobes. The phases actually received by the staves of the array are measured and deviations from the expected quadratic variation are determined and ascribed to constant interchannel phase errors. The measured errors include errors caused by stave

misalignment as well as electronic phase errors in the receiver
channels. Amplitude variations between the channels are also
determined during this calibration and are corrected by appropriately
adjusting the values.

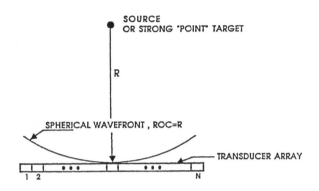

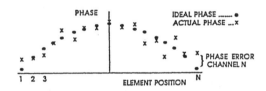

Figure 9.   Array phase calibration procedure.

This beamforming is illustrated diagrammatically in figure 10.
These new coefficient arrays are used to form beams in desired
directions (to cover an area to be searched for example). The power
and versatility of this beamforming approach is apparent, and as long
as the processor has sufficient reserve capability, all that is
required to form additional beams is the computation and storage of the
necessary beamformer coefficients.

3.5   Selection of Beamforming Coefficients

In the past, all multibeam side-scan sonars generated beams centered on
fixed array positions. The aperture length used to form a beam
increased with range (to maintain a constant linear-azimuth
resolution), but the beam array center remained fixed and did not vary
with range. The drawback of this technique (fixed beam array center)
is that at some ranges fairly large beam steering angles are required
for some of the beams. Since the level of the grating lobes (or
diffraction secondaries) in the beam pattern produced by a multi-stave
array depends upon the beam steering angles, and increases with

increasing steering angles, it is important to keep these steering
angles as small as possible.

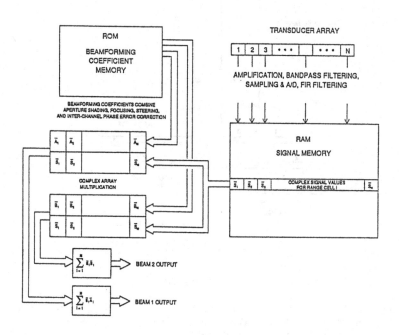

Figure 10.   Beamform processing.

With the digital beamforming approach described above, it is not
necessary to keep the beam array center fixed, and instead its position
can be varied with range to minimize the steering angle required to
form a given beam.   This process, varying the beam array center with
range, is illustrated in figure 11.   Basically, so long as is possible,
the stave nearest the center of the desired beam is selected as the
beam array center.   As the aperture length increases with range, it
becomes necessary to select beam array centers closer and closer to the
center of the array.   At the maximum range, all beam array centers
merge and the center of the array becomes the beam array center for all
beams.

An analysis must be carried out using a computer to determine the
optimum position for each beam array center for each range (for a given
tow speed).   The optimum position is the position that minimizes the
required steering angle for that range.   The payoff of this approach is
that the required beam steering angles, at all ranges, and for all
beams, are as small as possible and therefore the beam pattern grating
lobe levels are as slow as possible for the array used.

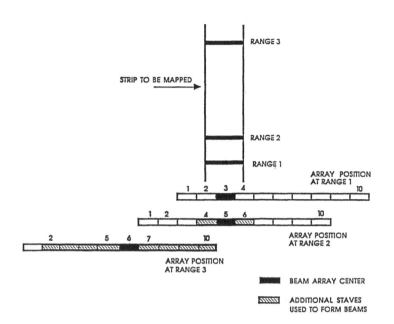

Figure 11.   Variable Beam Array center beamforming.

Although digital beamforming makes it possible to implement such techniques as the above optimum algorithm (variable beam array centers), it is important to point out that even without using such techniques the digital approach ensures better precision and better control of the beamforming process, and when properly applied will produce better beam patterns - beam patterns that will not get worse with use or with the aging of the sonar. Thus, digital implementation of current generation beamforming algorithms will produce better results than that provided by the analog hardware now used.

4.   NEW PERFORMANCE CAPABILITIES

One very important performance enhancement that can be provided easily in a digitally-beamformed sonar is the capability for simultaneous operation at several different frequencies. The sonar can operate at two or more nearby frequencies, and/or two or more widely separated frequencies. If the operation is at two nearby frequencies, pulses at each of the two frequencies are projected sequentially with a minimum inter-pulse delay (Fig. 12). Interference between the two frequencies is removed by filtering.

For simultaneous operation at two widely separated frequencies, the ping rate (pulse repetition rate) at the higher frequency is an integer multiple of the ping rate at the lower frequency. An example of the required pulse timing is shown in figure 12 for the case when the ping rate at the higher frequency is three times that at the lower frequency. For this timing scheme, the low frequency signal will not interfere with the high frequency signal; however, some degree of interference of the high frequency with low frequency will be experienced during the high frequency pulse transmission.

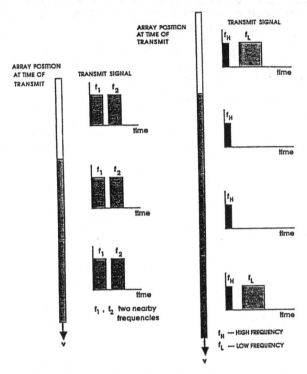

Figure 12. Simultaneous multi-frequency operation timing.

Simultaneous operation at more than one frequency requires only a modest increase in the amount of towfish hardware. Depending upon the power of the processor used and the processing requirements of the operation mode, an additional processor (or processors) may be required. In any case, however, the additional capability is provided with maximum ease since any additional hardware will be a duplicate (or duplicates) of the existing processing hardware, and only new beamforming coefficients must be provided.

Operation at two widely separated frequencies, for example 100 kHz and 600 kHz, can provide simultaneous long-range detection and short-range classification capabilities. For this example, the 100 kHz mode will operate out to 3 or 4 times the range of the 600 kHz mode. A single hydrophone array can be used, but separate projectors, operating at 100 kHz and 600 kHz, are required for proper insonification.

Simultaneous operation at two or more nearby frequencies will improve the performance of the sonar in terms of probability of detection and false alarm rate. Basically, the sonar forms the same beams at each frequency, a multi-look technique, and the outputs of the separate beams are summed incoherently to provide an output with improved quality. This multi-look technique increases the probability-of-detection performance and is particularly effective in improving the performance of long-range detection sonars.

ATS SYSTEM THEORY AND TEST RESULTS

ANTHONY ZAKNICH and PHILIP DOOLAN
Nautronix Ltd.
Fremantle
Western Australia

ABSTRACT. The technology and design methods behind commercial acoustic
positioning systems have not fundamentally changed over the last
decade, although some worthwhile improvements in performance have been
made. System designs have not given enough attention to the practical
limitations that the ocean presents to acoustic signals. Most of these
designs rely too much on the ocean being a homogeneous medium which
supports passage of undistorted straight line acoustic signals with
high signal to noise ratios. A new systems approach to the design of
acoustic positioning systems has been needed, which includes the
acoustic environment as a significant system design factor.
  The problem has been to develop a system which tracks beacons
accurately in the entire region below a single small hydrophone
assembly to usable ranges; minimizing the detrimental effects of
background acoustic noise, reverberation and multipathing effects on
system accuracy and reliability. The ATS acoustic positioning system,
developed recently by Nautronix Ltd. is the forerunner of a new
generation of such systems. Its design was spawned from the
recognition that the very successful ultrashort baseline systems had
significant operational advantages which were marred only by limited
accuracy and reliability in many practical working environments.
  The ATS system is a "very short baseline" acoustic positioning
system. Its hydrophones are mounted in a single Hydrophone Assembly
similar to an ultrashort baseline system, however its principle of
operation is different to the standard ultrashort baseline system.
Transponder beacon positions are determined by measuring slant range
and the direction of the beacon signal, using high resolution signal
detection and cross-correlation in conjunction with digital beamforming
techniques. The acoustic signal used is a train of short chirps in the
15 to 18 kHz frequency range. Current advances in electronics
technology has made it possible to develop this commercial acoustic
positioning system based on a growing body of underwater acoustics and
digital processing research.
  This paper presents the theory of operation of the ATS system,
showing its performance benefits supported by laboratory and field test
results. The test results show that the ATS system performs very
accurately and reliably in traditionally difficult acoustic

105

D. A. Ardus and M. A. Champ (eds.), Ocean Resources, Vol. II, 105–116.
© 1990 *Kluwer Academic Publishers. Printed in the Netherlands.*

environments such as long range in highly reverberant shallow waters
and when operating close to the water surface.

1. INTRODUCTION

This paper will examine the problems to be solved in designing and
evaluating an acoustic positioning system. It will also examine the
theory used in the design of the ATS acoustic positioning system and
provide a review of its operating characteristics and performance.

1.1 The System Design Criteria

A summary of the major system design requirements for a practical
acoustic positioning system are as follows:

(i)   The ability to track beacons accurately and reliably in the entire
      hemispherical region below a single Hydrophone Assembly.

(ii) A long range capability which points to the use of higher signal
      energies and lower signal frequencies having less attenuation in
      water.

(iii) High accuracy which, using standard ultrashort baseline phase
      measuring techniques, would require higher frequencies, giving
      shorter range capability.

(iv) Ability to work reliably and accurately at long ranges, in shallow
      waters and close to the water surface. This means facing the
      problems of reverberation and multipathing effects.

(v)   Ability to work reliably and accurately in high and varying
      background noise which, using ultrashort baseline methods, would
      require the ability to change operating frequencies to quieter
      frequency bands.

(vi) Easy to operate system with high quality graphic and numeric
      displays.

1.2 ATS's Solution to the Tracking Problem

The technology in ATS has been developed to specifically solve the main
acoustic problems which limit the accuracy and reliability of
conventional ultrashort baseline positioning systems. The development
of ATS has involved identifying these detrimental effects and solving
for them through the use of specialized signals and digital signal
processing technology.
     A summary of the ATS solution to the tracking problem is as
follows:

(i)   Use of a specially patterned medium frequency high energy signal
      based on linear frequency sweeps and matching filter techniques
      resulting in better signal detection and long range capability.

(ii) Use of a unique least squares estimate solution to an electronic
      beamforming algorithm in conjunction with a small, symmetric,
      redundant, plane hydrophone array, resulting in high accuracy and
      resolution of signal direction to minimize the multipathing
      problem in both deep and shallow waters.

(iii) Better graphic representation of positioning information using
      high resolution colour screens.

(iv) Very user friendly operating system based on dynamically allocated
      function keys in an ordered tree structure with on-screen help
      information available for every function in the system.

(v)   Large processing power and unique realization of digital signal
      processing techniques both in hardware and software making the
      solution to the problem commercially realizable in a compact
      package.

      The first production prototype of ATS was completed on 17th
December, 1987. Evaluation trials were conducted using various ships
at sea and on stable mountings across harbour facilities up to 13th
April, 1988. From then to the present the ATS system has been
demonstrated and used in a variety of real applications.

2.   ATS THEORETICAL BASIS OF OPERATION

The typical ultrashort baseline single frequency phase measuring system
assumes the following signal model:

$$r(t) = s(t) + N$$

where:
         $r(t)$ is the received signal
         $s(t)$ is the original transmitted signal
         N is the added transmission noise (assumed to be
                Gaussian).

      If the model is correct it is possible to filter the noise
effectively with a matching filter or narrow band-pass filter and
recover accurate and reliable signal phase information, so long as the
signal to noise ratio is high enough. However, this is not an adequate
model for real signals in the ocean, especially over long ranges, in
shallow waters and close to the water surface. The dominant system
noise is correlated noise rather than Gaussian distributed background
noise. The signal is dominated by multipathing effects caused by
micro-particle scatter, refraction, surface and bottom reflections.
This results in a high level of correlated noise content in the

received signal which makes it very difficult to accurately recover the
signal's phase information.
      A more suitable signal model adopted for ATS design is as follows:

          r(t) = s(t)*M(t) + N

where:
          M(t) is the multipathing effect which is very hard to define
          *   is the convolution operation.

      The multipathing factor M(t) is very difficult to define and there
is insufficient information in r(t) to find a unique solution for it.
To minimize the effects of M(t) on r(t) it is desirable to have a short
signal, but to maximize signal detection the signal should be long.
The classic phase measurement system uses signal frequencies of around
28 kHz and of approximately 1 ms duration as a compromise. This
compromise results in reduced range capability with a less than optimum
accuracy.
      The ATS technology improves on both range and accuracy by better
signal design and signal processing which minimizes the effects of
M(t). Firstly the signal is composed of a train of eight 1.2 ms chirps
(linear frequency sweeps) separated by 10 ms, in a lower frequency band
of around 16.5 kHz. Signal detection is improved because of increased
signal energy but the effects of M(t) are not increased. The lower
frequency and better signal detection give the better range capability.
As the noise effects of M(t) approximate to a normalized Gaussian
distribution, the angle bias due to M(t) is reduced and the angle
accuracy is improved. The ATS signal is longer with much more
frequency content so it is possible to exploit this effect of the
central limit theorem. Basically, this is done by the coherent
combination of the cross power spectrums of the eight chirps. These
cross spectral estimates are used in a least squares estimate
beamforming algorithm, done in the frequency domain, which solves for
the steering matrix which gives the LOOK direction. The least squares
estimate incorporates a total of 160 equations compiled from 10 unique
cross power spectrums by 16 frequency samples in each. All this
redundant information ensures a very good least squares estimate
solution for the LOOK direction as well as provides a quality factor
for the angle accuracy.
      The beamforming signal processing is done in the frequency domain
because it is the most logical and convenient way to do so. Classical
beamforming is done with a single frequency so it can be done easily in
the time domain. When the signal has significant frequency content it
becomes virtually impossible to model the phase or time delays which
result through the system electronics. Consequently, the hydrophones,
preamplifiers and filters require stable and precise calibration to
achieve high accuracy. The problem of long term calibration stability
is solved by using a symmetrical hydrophone array and by performing a
frequent self calibration process. A symmetric and redundant plane
array of 5 hydrophones is used which significantly reduces effects
detrimental to stability, after factory calibration, such as
temperature expansion and contraction and variation in sound velocity.

The system analog filters are calibrated for time delay drift regularly, with calibration signals similar to the received signals. This also allows the system to use much sharper analog band-pass filters to reduce more out-of-band noise without sacrificing accuracy due to long term phase drift.

Using chirps in the signal gives a number of significant benefits. Firstly, as the chirp has inherent information it is possible to use a spatially undersampled receiving array (i.e. separation of over half a wavelength). This bigger array gives better accuracy than a phased system using a single frequency which is forced to use a small array with a separation of less than half a wavelength (typically 3/8 wavelength). Secondly, a chirp can be detected more reliably than a single frequency burst using a matching filter.

The cost of using a train of chirps spread so far apart is that a velocity correction must be applied so that the chirps can be coherently combined. The received chirp train without noise components is as follows:

$$rc(t) = \sum_{i=0}^{7} c(t - iT - iTVrel/Vw)$$

where:

      rc(t) is the received chirp train
      T   is the chirp separation in time
      Vrel is the relative velocity of the beacon with respect
            to the hydrophone array
      Vw   is the velocity of sound in water.

The second term in c is the time delay factor of each chirp and the third term is the velocity correction factor. The ATS system solves for the velocity correction factor by assuming that the time separation between chirps is constant.

3.   ATS SYSTEM DESCRIPTION

The ATS-S02, ATS-S04 and ATS-S08 models are "very short baseline" acoustic positioning systems. Their receiving transducers are mounted in a single Hydrophone Assembly similar to an ultrashort baseline system, however the principle of operation is different.

The standard ATS system consists of:

-   Main Control Unit (19" rack mountable).

-   Hydrophone Assembly with integral Vertical Reference Unit plus 30 m of cable.

-   Vertical Reference Unit plus 30 m of cable.

-   Transponder Beacons (2, 4 or 8 depending on model).

The beacon positions are determined by measuring slant range and the precise direction of the transponder beacon signal. Transponder beacons are activated by a tone burst from the Hydrophone Assembly. After a fixed turn around time they return a specially patterned and coded signal to the Hydrophone Assembly.

The signal is received by five hydrophones, in the Hydrophone Assembly, forming a plane array with a very short baseline (separation of approximately one wave-length). The signals are hard limited after they are band-passed and amplified so only signal periods are used for digital processing. The five signals are run through buffers so that when signal detection and identification occurs, they can be captured for further processing. Digital signal detection is done in real-time at a lower sample rate than that which is subsequently applied to the captured signals.

The captured signals are transferred via a serial link to the main control unit where the positioning or tracking signal processing is performed. A least squares beamforming algorithm is applied which directly solves for the steering matrix giving the signal direction. The slant range to the beacon is simply determined by the signal travel time. Position corrections are made for the pitch and roll of the Hydrophone Assembly at the time of signal reception, thus giving the X, Y and Depth position coordinates of the beacon. The position coordinates may be displayed as raw calculated data or via a least squares estimate smoothing filter depending on the required application.

3.1 Transponder Mini-Beacon Signal Processing

The trigger signal to which a transponder beacon is tuned, is one of eight tone burst. The beacon receiver system uses a hard limiter amplifier and bandwidth ratio processor.

The beacon's transmission signal is a train of eight evenly spaced short linear frequency sweeps (chirps) in the 15 to 18 kHz band. The chirps are coded, adding more information for correct signal recovery. This train is followed by a short beacon identity tone burst (the same frequency as the beacon's interrogation frequency). Telemetry beacons have another train of four chirps following. The separation in time between the main eight chirps and the last four represents the telemetry value.

3.2 Hydrophone Assembly Signal Processing

The five received signals are filtered by a sharp 15 to 18 kHz bandwidth filter, hard limited, digitized and logged. The analog preamplifiers and filters are calibrated regularly by the system to measure the precise bulk time delays and frequency responses in each channel to ensure accurate signal time delay measurements. Real-time signal detection occurs in the centre channel using a hardware realized chirp matching filter.

After the signal is detected, the five raw signals which have been logged in buffers are saved for post processing in the Hydrophone

Assembly and then transferred to the Main Control Unit for final
processing.

The first post processing task is the determination of the first
chirp in the signal chirp train of the centre channel, to locate the
train precisely in the buffer. Velocity correction is done using the
fact that the eight chirps are equally spaced in time. Velocity
correction is effective over 0 to ±10 m/s relative motion.

The logged signals are transferred to the Main Control Unit
processor via a high speed RS422 serial link. All communication with
the Hydrophone Assembly is via this single two way serial link. The
data is encoded with full error checking codes, ensuring accurate
transfer over very long cable lengths.

The Hydrophone has a single transmitting transducer which
transmits tone burst triggers for beacons. The tones are approximately
5 ms long and are one of eight frequencies equally spaced in the 15 to
18 kHz band.

3.3 Main Control Unit Signal Processing

The raw signal data transferred from the Hydrophone Assembly is turned
into coherently combined cross power spectrums. These are used in a
single least squares estimate beamforming algorithm, done in the
frequency domain, which solves directly for the steering matrix. The
steering matrix is relatively independent of signal amplitude, which is
why the signal amplitude is discarded earlier on to save processing
time. This simplifies the digital processing requirements and allows
tracking position calculation to be done in approximately one second
using one National Semiconductor 32032 CPU running at 10 MHz and one
Intel 80186 CPU running at 8 MHz. After the tracking position
calculation is done, the Hydrophone Assembly pitch and roll angle
corrections are applied to produce the final beacon position
coordinate.

4. ACOUSTIC POSITIONING SYSTEM PERFORMANCE EVALUATION

The performance evaluation of a very accurate acoustic positioning
system such as the ATS system requires a careful test set up and
analysis of the data. To be meaningful, the analysis must be done on a
large sample of raw unfiltered data. Any filtering, unless it is
specifically designed with full knowledge of the data's statistics,
will introduce bias to the results and prevent a true performance
evaluation. For accurate results, filtering is most useful only after
all the data has been gathered, statistically analyzed and integrated
with external position information such as from RF navigation
equipment. Real-time filtering of tracking data is of limited value
except to give a smoother display for an operator's benefit.

A system's accuracy is most conveniently expressed in terms of
bearing angle error, i.e. bearing angle in the horizontal plane. To
measure it, the Hydrophone Assembly must be mounted vertically on a
shaft and be held rigid and stable in a deep body of water with a fixed
beacon a suitable distance away. The Hydrophone Assembly must also be

able to be rotated and measured accurately via the shaft rotation in
azimuth angle. With this test set up it is possible to test the
system's repeatability as well as relative and absolute accuracies for
different beacon depths. The tests involve the gathering of the solved
X, Y position records, where each record should have at least 100
samples or more to establish statistically significant conclusions.
     The bearing angle is determined from the solved X, Y horizontal
position coordinates as follows:

$$A = arcTan(Y/X)$$

where:
     A is the solved bearing angle (expressed in degrees).

     The relationship between bearing angle error and the error in X or
Y can be expressed as a differential equation:

$$ds = r\,dA$$

where:
     ds is the error in X or Y
     r is the horizontal range
     dA is the error in bearing angle.

A convenient derivation of this formula is as follows:

% error in X or Y = 1.745 x (error in A in degrees)

4.1 System Repeatability

Repeatability is a measure of the system's ability to reproduce a given
angle solution. A record of position samples is taken at a specific
bearing angle and beacon depth and its mean and RMS error are
calculated as follows:

$$\text{Mean } A = \sum_{i=1}^{N}(Ai)/N, \text{ and}$$

$$\text{RMS error in } A = SQRT((\sum_{i=1}^{N} SQR(Ai - \text{Mean } A))/N)$$

where:
     N is the number of samples in the record
     Ai is the ith angle sample.

     The mean represents the solved bearing angle whilst the RMS error
in A shows the repeatability. The repeatability figure includes both
angle bias and random jitter errors. The angle bias is a result of
signal corruption which the system signal processing cannot eliminate.
If this test is done with the beacon very close in a stable and deep

body of water, the RMS error is mainly due to the random jitter and represents the best repeatability the system can achieve.

4.2 System Accuracy

Accuracy in bearing angle is the measure of the system's ability to solve for the correct angle that the beacon is at with respect to the Hydrophone Assembly. To determine relative and absolute accuracies the Hydrophone Assembly is rotated to a series of known and evenly spaced bearing angles over the full 360° range. At each angle, records are taken, and the mean A's are calculated. The RMS error of each record gives the repeatability as before. The error associated with each angle is determined as follows:

$$\text{Angle Error}_j = (\text{Solved Angle}_j - \text{Shaft Angle}_j - \text{Reference Angle})$$

where:

Solved Angle$_j$ is the mean A at the jth angle as we go from 0 to 360°.

Shaft Angle$_j$ is the angle to which the Hydrophone has been rotated.

Reference Angle is the offset between the mechanical and acoustic zero.

If we take the mean of the Angle Errors this gives us the relative angle bias which is composed of a combination of signal corruption and calibration factors. If the test is done with the beacon very close in a stable and deep body of water, the bias is mainly due to calibration factors. The RMS error of the mean gives the relative system accuracy. The absolute system accuracy is determined by assuming that the mean of the Angle Errors is zero and calculating the RMS error as follows:

$$\text{RMS error} = \text{SQRT}((\sum_{j=1}^{N} \text{SQR}(\text{Angle Error}_j - 0))/N)$$

ATS system accuracy figures achieved during calibration in a test tank at approximately 1.5 m range are as follows. Relative bearing angle accuracies range between 0.061 to 0.380 RMS error in degrees with a mean of 0.141 degrees for beacon angles of -5 to 70 degrees down from the horizontal for the full 360 degree bearing angle ranges. For declination angles between -5 to 45 degrees, the relative accuracies are much better, ranging between 0.061 to 0.135 with a mean of 0.096 degrees. The absolute bearing accuracies are very similar, ranging between 0.065 to 0.492 with a mean of 0.174 degrees for declination angles between -5 to 70 degrees. For declination angles between -5 to

45 degrees, the absolute bearing accuracies range between 0.0650 to
0.121 with a mean of 0.101 degrees.

5.   ATS TEST AND APPLICATION SUMMARIES

The following ATS system performance specifications have been
summarized from extensive factory and field tests as well as
demonstrations to prominent customers.

1.   Under conditions with negligible environmental corruptions, such
     as in stable and deep waters, ATS is capable of relative bearing
     angle accuracy of 0.144 degrees RMS error (i.e. 0.25% of slant
     range in horizontal X and Y position coordinates).

2.   Under harsh environmental conditions, such as very shallow waters,
     ATS is able to produce relative bearing angle accuracies of better
     than 0.37 degrees RMS error (i.e. 0.65% of slant range in
     horizontal X and Y position coordinates) out to several hundred
     metres. An example environment is a relatively quiet lake with
     water depths of 4 to 8 m to a range of 700 m.

3.   Under extremely harsh environmental conditions such as noisy, very
     shallow and unstable waters, ATS is able to produce bearing angle
     accuracies of better than 0.6 degrees RMS error (i.e. 1.0% of
     slant range in horizontal X and Y position coordinates) out to
     several hundred metres. An example environment is across a busy
     harbour mouth with strong tide flowing, turbulent waters with a
     maximum depth of 11 m out to a range of 240 m.

     Tracking Range is dependent on many factors including the level of
background noise, but the standard ATS system can be stated to have a
minimum range capability of at least 1000+ m in typical working
conditions with a capacity to be useful up to at least 3000+ m. The
ATS system, using a transponder Mini-Beacon is capable of relative
slant range RMS errors of as low as 0.06 m depending on the stability
and homogeneity of the water.
     Under the very worst environmental conditions where tracking
systems must operate, the ATS system is able to return accuracies which
are equal to or better than that which ultrashort baseline systems can
achieve under favourable conditions. In other less harsh environments,
the ATS system can return much better accuracies and much more reliable
operation. Its usable range is greater than that of other ultrashort
baseline systems.

5.1 Application Areas

A summary of the areas and applications in which the ATS system has
been successfully trialled to date are as follows:

1.   Tracking in shallow water depths of between 4 m to 20 m close to the water surface and to maximum ranges up to 4000 m.

2.   Instrumented towed array tracking in deep waters to ranges of 500 m and 1000 m with beacon depths of between 10 m to 100 m.

3.   Geophysical seismic streamer tracking.

4.   Sidescan Towfish tracking.

5.   Remotely Operated Vehicle (ROV) tracking.

6.   ROV tracking in and around platforms for platform cleaning.

There are still more areas of application yet to be verified, but enough have been tried to show that ATS will perform very well in traditional tracking applications as well as opening up new areas not possible before.
The following is a list of major institutions and companies which have satisfactorily used or trialled the ATS system to date.

1.   Woodside Offshore Petroleum Pty Ltd.: Demonstration on 2-27 May 1988 on M.V. Shelf Supporter - ROV tracking for platform cleaning off the Northwest Shelf of Western Australia.

2.   Martin Marietta Baltimore Aerospace: Purchase for ROV tracking.

3.   The Applied Physics Laboratory (APL) of the University of Washington, Seattle: Purchased and used September 1988 on R.V. Knorr - instrumented towed array tracking in Greenland sea.

4.   Geophysical Services Inc.: Demonstration on 14-19 October 1988 on M.V. Pacific Titan - seismic streamer tracking in Timor sea.

5.   Harvey-Lynch Inc.: Purchase - seismic streamer tracking in Gulf of Mexico.

6.   Swedish Navy: Demonstration on 8 November 1988 in Stockholm harbour.

7.   Gardline Surveys: Demonstration on 20 November 1988 on M.V. Sea Searcher - sidescan towfish and other towed object tracking to 1200 m ranges in the North sea.

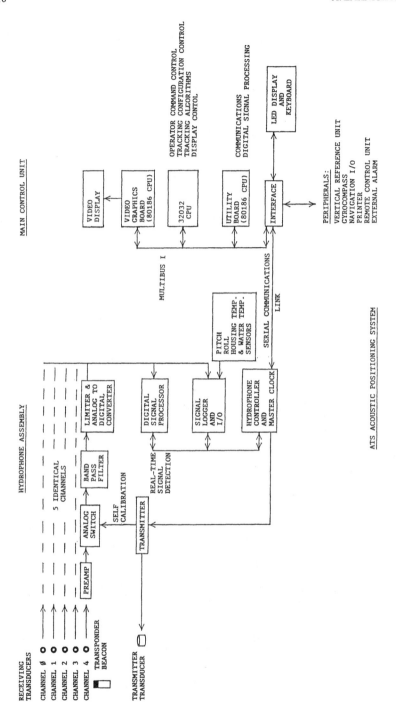

Figure 1.   ATS acoustic positioning system

LONG AND SHORT RANGE MULTIPATH PROPAGATION MODELS FOR USE IN THE
TRANSMISSION OF HIGH-SPEED DATA THROUGH UNDERWATER CHANNELS

A. FALAHATI, S.C. BATEMAN and B. WOODWARD
Department of Electronic & Electrical Engineering
University of Technology
Loughborough, Leicestershire, LE11 3TU
U.K.

ABSTRACT. There are many applications requiring the propagation of
acoustic signals through an underwater channel from a mobile
transmitter to a mobile receiver, for example from a diver or
autonomous vehicle to a surface vessel, or vice versa. For a diver,
the information can represent depth or physiological parameters such as
heart rate, breathing rate and body temperature. For an autonomous
vehicle it can represent depth, control commands or scientific data.
During transmission, the signals can be corrupted by noise from many
sources. They can also be reflected and scattered at the surface, and
refracted by variations in acoustic velocity; these effects cause
multipath interference at a receiver. Here, we describe a diving
application to illustrate design problems associated with a low-rate
telemetry system. Then we consider the constraints of extending it to
a high-rate system, including effects limiting its range.

1. INTRODUCTION

At Loughborough University we have specialised in the design of sonar
systems, transducer arrays and underwater position fixing systems. We
have also studied the problems of underwater communications and data
telemetry in relation to diving applications (see Reference section).
In this paper we discuss the design of underwater acoustic systems and
their limitations in the context of a diver monitoring system.
Although this is a specific application, the principles are general and
the parameters can be adapted for other applications.
        There is some dispute about the rationale of diver monitoring,
especially for commercial divers who do demanding work at great depths.
In the interests of safety, it might seem justifiable to monitor
divers' physiological well-being, but many object to being instrumented
during operational diving.
        Monitoring can, however, fulfil a useful role for research
purposes among military, scientific and amateur divers. For example,
the Royal Navy's diving vessel HMS Challenger has facilities for
physiological monitoring of two divers working from a bell to a depth
of 300 m. The US National Plan for Safety and Health of Divers also

117

D. A. Ardus and M. A. Champ (eds.), Ocean Resources, Vol. II, 117–124.
© 1990 Kluwer Academic Publishers. Printed in the Netherlands.

covers physiological monitoring, including inspired oxygen partial
pressure, expired carbon dioxide partial pressure, and deep body (core)
temperature (Pearson, 1981) (Bannister). At Loughborough University we
are tape recording heart rates of sports divers in a joint study with
the UK Health and Safety Executive.

For any trials like these the diver's transmitter unit must be
small enough to be carried without encumbrance, for example in a
waterproof housing strapped to his breathing gas cylinder or under his
drysuit. The sensors must be easily attached and comfortable to wear
without affecting movements.

2.   SYSTEM DESIGN

One of our systems is designed to monitor heart rate, breathing rate,
temperature and depth once very second over a range of 300 m in the
presence of echoes and noise. The four sets of data are transmitted as
8-bit words; thus with an additional synchronising word the system
operates at 40 bits per second. This slow rate is adequate because
none of the parameters is likely to change very much in one second.
The transmission format allows comparatively long periods between the
data words.

The basic transmission arrangement, shown in figure 1, consists of
a projector P for transmitting the data from the diver and a hydrophone
H for receiving it remotely. For the best signal-to-noise ratio at the
receiver these should be a matched pair of piezoelectric transducers.
If they have omnidirectional characteristics there is no need for
accurate alignment.

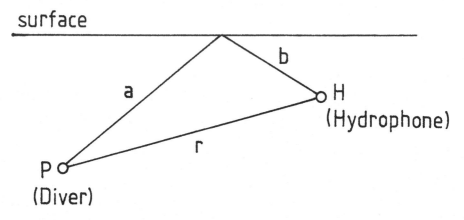

Figure 1. Multipath interference.

With these transducers, data symbols can only be transmitted
through water by modulating a carrier frequency so that each symbol is
sent as a tone burst. There is no mandatory allocation of carrier
frequencies for underwater acoustic systems (e.g. side scan, sector
scan, doppler log, correlation log, profiling sonar, beacons,

transponders, telephones, etc.). A suitable choice for this
application is 75 kHz because it is a compromise between the size and
cost of transducers and the effects of attenuation and noise. It is
also well above the recommended emergency frequencies (in the UK) for
manned submersible beacons (10 kHz) through-water speech (8-11 kHz),
hazard makers (13 kHz) and diving bells (37.5 kHz).

The modulation method affects the complexity of the transmitter
and its performance in a noisy environment like the sea. The three
main methods are amplitude shift keying (ASK), phase shift keying (PSK)
and frequency shift keying (FSK). FSK is widely used in underwater
applications because, as with PSK but unlike ASK, both binary states
are characterised by the presence of a carrier: f_0 for binary '0' and

f_1 for binary '1'. Also, an FSK receiver with narrow bandwidth filters
feeding into a signal level comparator offers good echo rejection since
the direct signal level is higher than that of the echo signal.

Typically, FSK has bit error rates of 10^{-5}, but this depends on the
signal-to-noise ratio. Although two frequencies are needed for FSK,
they can be transmitted with one transducer operating either side of
its resonant frequency.

One of the main problems is the corruption of data due to
multipath interference. Figure 1 shows the problem for a single
surface reflection. If the acoustic velocity is v m/s, a burst of data
from P arrives at H via the direct path after r/v s, and via the echo
path after (a+b)/v s. A simple receiver can therefore detect a
duplicate set of data symbols, assuming the echo level is high enough.
For multiple echoes, the received data can overlap in time and become
degraded.

Echo interference can be largely avoided by using asynchronous
serial transmission. Thus, the transmitter output consists of time
division multiplexed (TDM) bursts of data, each followed by a quiet
period. The transmission timing format of our diver monitoring system
is shown in figure 2. Each of the four 8-bit data words is transmitted
as a 2 ms group of symbol bursts. Allowing 50 µs between each burst to
ensure adequate reduction of waveform transients fixes the burst
duration at 200 µs. With a synchronising channel included for timing,
there are five channels in a 1 s frame, and each channel is transmitted
in 200 ms. This format avoids multipath interference for path length
differences of between about 3 m and 300 m. Differences of less than
3 m cause a time-delayed version of each 2 ms group to be received
simultaneously with the group arriving via the direct path.
Differences of greater than 297 m result in interburst interference,
for example channel 1 and channel 2 may arrive simultaneously

The effect of interference depends upon the relative levels of the
direct and reflected components and the discrimination of the receiver.
The short duration of the burst and the long period between bursts
allows all echoes to reach H before the next burst leaves P. After the
arrival of a burst via the direct path the receiver is muted to reject
echoes as spurious data.

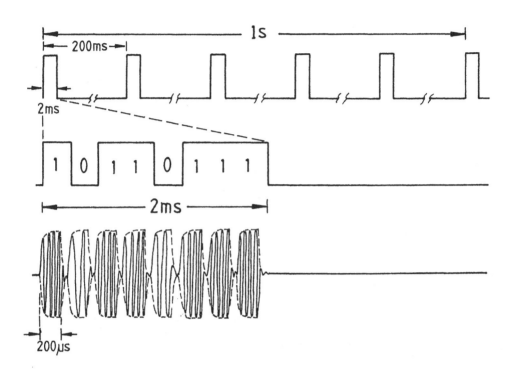

Figure 2.   Transmission timing format.

    While the design of an asynchronous low data-rate system is
relatively straightforward, there are far more problems in designing a
synchronous high data-rate system, especially over long transmission
paths.   In future developments, the quiet periods between bursts can be
used for the transmission of speech, position coordinates, and perhaps
scientific data, i.e. transmitting a continuous stream of data
comprising data sets of differing durations.   Depending on the symbol
rate and relative path lengths, the first-arrived data symbols reaching
H (Fig. 1) may be detected correctly.   But after a delay, the echo
stream will also reach H, resulting in constructive or destructive
interference.   If the diver is moving, the path lengths are
continuously changing and so interference will occur intermittently.
Consequently, the amplitude and phase of the received symbols will vary
with time.   Indeed, the amplitude variations may be very severe,
causing the received signal strength to reduce by as much as 40 dB, an
effect known as signal (flat) fading.

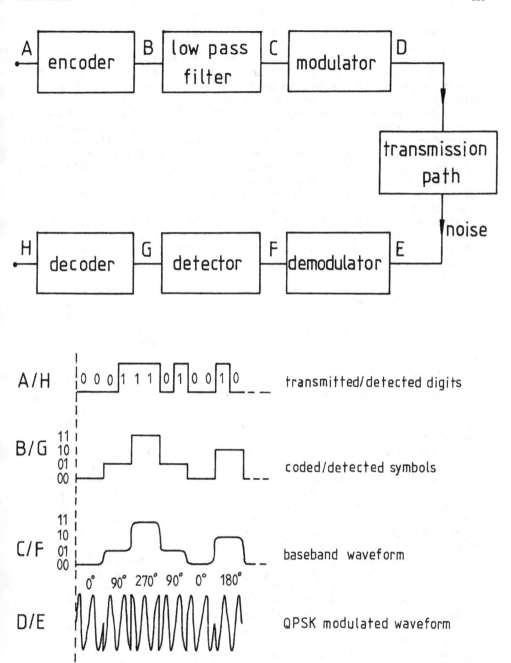

Figure 3. Model of data transmission system.

3. MODEL OF AN UNDERWATER MULTIPATH CHANNEL

A digital data transmission system can be represented by the model
shown in figure 3. The transmitter and receiver sections are typical
of most communication system models, but the transmission path is quite
different for the underwater acoustic case.
   The transmitter comprises an encoder, a low pass filter and a
modulator. The input to the encoder is a stream of binary digits
representing the information to be transmitted. These bits may be sent
from a multiplexer that selects the parallel outputs from several
sensors in turn and converts them into serial form. A suitable bit
rate for synchronous transmission underwater would be 2400 bits per
second, i.e. each bit has a duration T of about 417 µs. Fourier
analysis shows that a continuous stream of alternate 1's and 0's
requires a wider bandwidth than any other combination of bits. Since
there is a finite probability of this occuring, at least for a short
time, we must satisfy the bandwidth constraints for this 'worst' case.
If the maximum bit rate is 2400, the corresponding minimum bandwidth is
1200 Hz.
   The encoder samples successive pairs of bits and generates a 4-
level stream of symbols. Each level corresponds to a different bit
combination, e.g. 00, 01, 10, 11. Since each level has a duration of
2T seconds, the symbol rate is half the bit rate and the bandwidth
requires is therefore halved. In our practical example the bandwidth
is 600 Hz.
   By passing the 4-level coded symbols through a low-pass filter,
the high frequency harmonics of each symbol are suppressed, and the
bandwidth is therefore further reduced. The filter is essentially a
waveform shaper that produces a 4-level shaped baseband waveform; it is
this waveform that modulates the carrier waveform for transmission
through the water.
   Modulation here is by quadrature phase shift keying (QPSK), so
that each of the four levels in the baseband waveform produces a phase
change of $0°$, $90°$, $180°$ or $270°$. These absolute phase changes can be
generated by the modulator, but because of the phase changes that
inevitably occur in the underwater channel it is preferable if it
generates differential phase modulation. This means the receiver has
only to detect a phase change between any particular symbol and the
preceding one rather than the first one in a long stream.
   As mentioned earlier, various effects lead to multipath
propagation of the acoustic waves in the transmission path. One of
these is due to the variation of the acoustic velocity from layer to
layer due to changes in temperature, salinity and pressure. This
produces small continuous variations in the acoustic refractive index,
and hence affects the direction of an acoustic ray. Another cause is
reflection and scattering at the surface. This generates additional
rays which are detected with random phase relationships. Also, the
modulated carrier signals are degraded by added noise.
   At the receiver these signals are demodulated, detected and
decoded. The output of the demodulator is a noisy and distorted 4-
level baseband signal. In principle, it is a noisy version of the
waveform that was applied to the modulator in the transmitter. The

detector (a non-linear equaliser) then shapes this waveform to produce
properly 'squared' 4-level detected symbols. Finally, the decoder
generates a stream of bits; if there are no errors these will represent
the original data that was transmitted.

4.   DISCUSSION

In order to simulate the multipath effects in an underwater channel,
two computer models have been devised, one for short-range and one for
long-range transmission distances. In the short-range model, the
acoustic rays from the transmitter are treated independently since, at
any instant, the receiver could be in the vicinity of any one ray out
of several that will be travelling over different paths. However, for
the long-range model, the received signal is treated as the
superposition of a number of time-delayed components arriving over
different paths. Both methods model the time-varying characteristics
of channels observed in practice. They are being tested in full system
simulations to predict the error rate performance of different
modulation methods and detection techniques. If predicted bit error
rates are sufficiently low, a practical system with a high data-rate
can be implemented.

5.   REFERENCES

Bannister, L.R. 'A diving mission recorder', ibid.

Bateman, S.C. (1988) 'The design and implementation of high-speed
    digital data communication systems', Int. C.I.S. J., Vol. 2,
    No. 3.

Brock, D.C., Bateman, S.C. and Woodward, B. (1986) 'Underwater
    acoustic transmission of low-rate digital data', Ultrasonics,
    Vol. 24, No. 4, pp.183-188.

Falahati, A., Bateman, S.C. and Woodward, B. (1987) 'High speed
    digital data transmission in an underwater channel', Proc. Inst.
    Acoust., Vol. 9, Part 4, pp.28-35.

Falahati, A., Bateman, S.C. and Woodward, B. (1988) 'Outline of a
    digital modem for an underwater communication channel', Fifth
    Int. Conf.: Digital Processing of Signals in Communications,
    Loughborough, IEE Publication 82.

Hardman, P.A. and Woodward, B. (1984) 'Underwater location fixing
    by a diver-operated acoustic telemetry system', Acustica,
    Vol. 55, No. 1, pp.34-44.

Hodder, T.M. and Woodward, B. (1986) 'Algorithms for underwater
    position fixing', Int. J. Math. Educ. Sci. Tech., Vol. 17,
    No. 4, pp.407-417.

Pearson, R.R. (1981) 'Why do we need diver monitoring?', Int.
Conf. 'Divetech' 81: The Way Ahead in Diving Technology, Society
for Underwater Technology, London.

Woodward, B. and Warnes, L.A.A. (1980) 'Signal processing
considerations in the design of a sonar system for divers',
Proc. Inst. Acoust: Signal Processing in Underwater Acoustics,
Loughborough University of Technology.

Woodward, B. (1982) 'Microprocessor-controlled diver navigation',
Int. Underwater Syst. Design, Vol. 4, No. 2, pp.10-15.

Woodward, B. (1985) 'A self-calibrating diver's rangefinder',
J. Acoust. Soc. Am., Vol. 77, No. 3, pp.1000-1002.

FEATURES OF BEAMSTEERING AND EQUALIZATION WHEN APPLIED TO HYDROACOUSTIC COMMUNICATION*

GEIR HELGE SANDSMARK
Division of Telecommunication
Norwegian Institute of Technology
Trondheim
Norway

1. INTRODUCTION

Recent development of tetherless ROVs motivates for wireless communication between the ROV and a surface vessel. For physical reasons, the use of hydroacoustic waves seems to be the most promising way to meet practical range requirements (500-1000 m). The most challenging application for hydroacoustic data transmission is transmission of images from a video camera or high resolution sonar as the required bit rate for such transmission is rather high (>10 kbits/s). Multipath interference due to reflections from the sea surface, the bottom or obstacles within the insonified water volume is likely to corrupt most attempts for hydroacoustic data transmission.

In Section 2 of this paper novel results are presented, describing how adaptive beamforming and adaptive equalization complements each other in combatting multipath interference, thus fitting well in a combined system.

Section 3 contains results describing the performance of the stochastic gradient lattice equalizer operating in a stationary environment as well as a time-variant case.

2. THE IMPACT OF ADAPTIVE BEAMFORMING AND ADAPTIVE EQUALIZATION ON THE CHANNEL IMPULSE RESPONSE

2.1 Beamforming/Beamsteering

As described in the previous section, reflections from surface, bottom and in-sea obstacles may cause the transmitted signal to reach the receiver via different paths. These rays will usually approach the receiver from different angles. It is therefore possible to attenuate some of these unwanted signals by applying a narrow transducer beam directed towards the direct path.

* This work was financed by NTNF OT.25.18524 and carried out in cooperation with SIMRAD Subsea A/S.

D. A. Ardus and M. A. Champ (eds.), Ocean Resources, Vol. II, 125–133.
© 1990 Kluwer Academic Publishers. Printed in the Netherlands.

When applying a simple tracking algorithm (Collins et al, 1985)
(Roberts, 1983) together with traditional beamforming (Monzingo and
Miller, 1980) unwanted signals (including multipath signals and noise
sources) outside the main lobe are attenuated according to the side-
lobe level at the actual angle of incidence. Further suppression is
possible by using Widrow's side-lobe canceller (Widrow and Stearns,
1985) or an optimum beamforming method (Monzingo and Miller, 1980)
(Haykin, 1985) (Haykin, 1986).
     An interesting aspect of the beamforming technique is revealed by
observing a relation between incidence angle and propagation delay.
This is illustrated in figure 1.

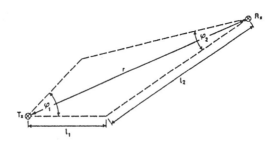

Figure 1. Geometry of area covered by transmitter and receiver
mainlobes.

     The longest possible path from transmitter to receiver via one
single reflector within the joint area is the route via one of the area
borders. Consequently, the longest delay relative to the direct path
depends heavily upon receiver and transmitter beamwidths. By
simplifying the channel impulse response to consist only of the direct
path together with single reflector paths within the joint area, this
also gives a limit of impulse response duration.
     From the figure we calculate the delay difference between the
direct path and the maximum delayed path as

$$\Delta T_{max} = \frac{1_1 + 1_2 - r}{c} = \frac{r}{c} \; \frac{\sin \frac{\psi_1}{2} + \sin \frac{\psi_2}{2} - \sin(\frac{\psi_1 + \psi_2}{2})}{\sin(\frac{\psi_1 + \psi_2}{2})} \qquad (1)$$

where c is the sound speed in sea water.

     Figure 2 depicts the maximum delay difference as a function of
beamwidths at 500 m distance between the transducers.
     As an example we chose 60° beamwidth for the transmitter and 5°
beamwidth for the receiver. From the diagram it is found that a
maximum multipath delay of 4 ms is still possible. As this corresponds
to 40 symbol intervals at 10 kHz symbol rate, severe intersymbol
interference is likely to corrupt the received signal. This motivates
the use of an adaptive equalizer together with the transducer system.

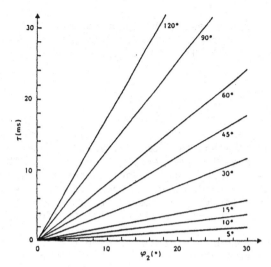

Figure 2. Delay difference as a function of transducer beamwidth.

2.2 Adaptive Equalization

While the narrow beam transducer prevents intersymbol interference by
attenuation of auxiliary paths, the equalizer attempts to compensate
for the resulting channel transfer function.
    Assume the following model of the received signal

$$y(t) = \underline{h}_N^H \, \underline{d}_N(t) + n(t) \qquad\qquad (2)$$

where $y(t)$ is the received signal sample at time t, $\underline{h}_N$ is the nx1
channel impulse response vector, and $\underline{d}_N(t)$ is the vector of the N last
transmitted data symbols, $n(t)$ is an additive white noise term.
Superscript H means Hermitian transponse.
    The scope of the linear adaptive equalizer is to adjust the
impulse response $\underline{W}$ of the FIR filter to minimize the power of the
difference between the filter output $d(t)$ and the desired signal $d(t)$.
This is achieved by calculating $\underline{W}$ according to the normal equation,

$$\underline{W} = R^{-1}p \qquad\qquad (3)$$

where R denotes the LxL autocorrelation matrix of the received signal
y, and p is the Lx1 vector describing the cross-correlation between the
desired signal d and the received signal y.
    For details on linear equalization refer (Haykin, 1986) or
(Proakis, 1983). R and p depend on the channel impulse response $\underline{h}$.
The autocorrelation matrix R is also affected by the noise power.
Detailed descriptions of these interrelations are given in (Sandsmark
and Solstad, 1987).

The simulations described in this paper are conducted using a
variant of the stochastic gradient lattice algorithm (Proakis, 1983)
(Honig and Messerschmitt, 1984).

3.   SIMULATION RESULTS

3.1 Stationary Case

We shall now examine the results from a simulation example. The
horizontal distance is varied between 100 m and 200 m. This variation
will only affect the impulse response and not the signal to noise ratio
as the channel simulator produces noise free signals to which noise is
added afterwards. Similarly, the receiver beamwidth is varied between
5° and 15°. Also this variation is done without affecting the signal
to noise ratio.
     The transmitter beamwidth and center direction are chosen in a
manner which sites both the direct path and the bottom reflection
within the transmitter transducer main lobe. This is done to produce a
"difficult" channel. The impulse response is calculated by a ray
tracing model including refraction and boundary roughness.
     Figure 3 shows the resulting impulse response at 150 m horizontal
distance. The receiver beamwidth does not affect the channel very
much. This is due to the special geometry. That is, there are no
significant signal contributions approaching the receiver in the
intervals between ±2.5° to ±7.5° relative to the beam axis. The strong
bottom reflection thus reaches the receiver at an incident angle less
than 2.5° away from the direct signal path and delayed about 1 ms.
     Figure 4 shows the learning characteristics of the stochastic
gradient lattice equalizer (Proakis, 1983) (Honig and Messerschmitt,
1984). The upper curve is obtained for input S/N = 10 dB, while the
lower curve fits S/N = 20 dB.

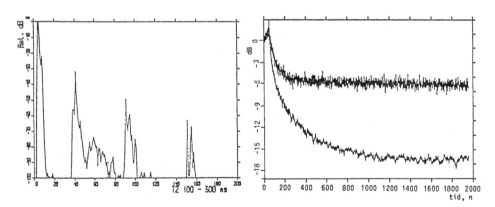

Figure 3. Resulting impulse
response at 150 m horizontal
distance and receiver beamwidth
15°. Bandwidth 1 kHz.

Figure 4. Learning characteristics
of the stochastic gradient lattice
equalizer.

Table 1 presents detection performance of the receiver with and without the adaptive equalizer. We notice that the error rates obtained from <u>noise-free</u> transmission through the channel are unacceptable for communication. On the contrary, by adding the adaptive equalizer to the system, the bit error rate becomes considerably lower and allows for reliable transmission. This confirms that even with a narrow beam (5°) receiver transducer multipath propagation may introduce severe intersymbol interference. Further it is shown that by including adaptive equalization reliable communication is still achievable.

TABLE 1. Error rate measured for different channels.

(a) Noise-free transmission without equalizer.

Horizontal Distance	Receiver Beamwidth	Bit Error Rate
150 m	15°	0.25
150 m	10°	0.25
200 m	10°	0.12

(b) Simulations using equalizer

Horizontal Distance	Receiver Beamwidth	S/N	Bit Error Rate
150 m	15°	20 dB	$1.6 \cdot 10^{-4}$
150 m	10°	20 dB	$1.6 \cdot 10^{-4}$
200 m	10°	10 dB	0.0046
200 m	10°	20 dB	$1.6 \cdot 10^{-4}$
200 m	10°	30 dB	0

3.2 Time-Variant Case

In this case we have used the discrete multipath model (4) to represent the channel

$$h(n,t) = \delta(n) + \sum_{i=1}^{L} \delta(n-h_i)a_i e^{-j\psi_i(t)} \qquad (4)$$

From figure 5 we now depict the variation of h(n) taking into account a linear motion of the receiver or the transmitter platform. The amplitudes a_i are approximately constant for small movements

relative to the distance between the receiver and the transmitter. The amplitudes depend on the path lengths and the path locations within the directivity patterns of the transducers. For time intervals of about 2.0 s, velocities less than 0.5 m/s and distances greater than 50 m the constant amplitude assumption is a good approximation for cases when the path is not too close to the zeroes of the directivity patterns. The delay n_i and the phase shift ψ_i are both dependent upon the

difference between the actual path distance and the distance of the direct path. For narrow-bandwidth signals n_i is approximately constant

for small movements. Consequently, only the variation of the phase ψ_i

is taken into account.

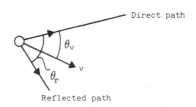

Figure 5. Geometry for channel model derivation.

The phase variation vs. time of each path is due to the doppler-shift.
Now, assuming that the carrier recovery loop is compensating for the doppler shift of the direct path, the phase ψ_i of the i'th

reflected path ψ percepted by the equalizer, becomes

$$\psi_i(t) = 2\pi \frac{f_o}{f_s} \frac{v}{c_o}[\cos(\theta_v - \theta_r) - \cos\theta_v]t \qquad (5)$$

f_s is the sampling frequency, f_o is the carrier frequency, and c_o is the speed of sound.

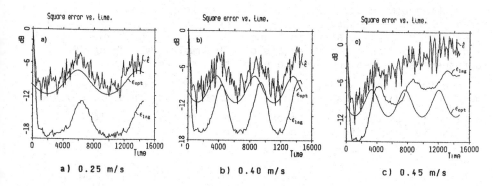

a) 0.25 m/s   b) 0.40 m/s   c) 0.45 m/s

Figure 6.   Equalizer error performance for different velocities.

Figure 6 depicts the tracking performance of the stochastic gradient lattice algorithm obtained for three different velocities. The different measures of error are defined in Equations 6 through 8.

$$\varepsilon(t) = \frac{1}{10} \sum_{j=1}^{10} |e_j(t)|^2 \tag{6}$$

where $e_j(t)$ is the error of the equalizer output at time t for the j'th trial.

$$\varepsilon_{opt}(t) = 1 - \underline{W}_{opt}^{H}(t)\underline{p}(t) \tag{7}$$

where $p(t)$ is the cross-correlation vector described in Section 2.2, and $\underline{W}_{opt}$ is the optimum equalizer impulse response calculated from (3).

$$\varepsilon_{lag}(t) = [\underline{W}_{opt}(t) - \underline{\hat{W}}(t)]^{H} R[\underline{W}_{opt}(t) - \hat{W}(t)] \tag{8}$$

where $\hat{W}(t)$ is the actual equalizer impulse response at time t averaged over 10 independent sequences, ε is an estimate of the mean square error, ε_{opt} denotes the theoretical limit of equalizer performance while ε_{lag} is the increase in output error due to equalizer misadjustment (Haykin, 1986) (Honig and Messerschmitt, 1984).

From the figure we can see that for low velocities the estimated mean square error approximates its optimum value, and correspondingly, the lag error is small. For higher velocities the lag error grows thereby causing significant increase in the measured error. The severe problem associated with the large lag error is the increase in decision errors. Since the detected symbols are used as a reference for the updating algorithm, the decision errors corrupt the reference, thus

preventing proper updating. The result is loss of track as
demonstrated in example 6(c). A measure of error probability as a
function of time, is shown in figure 7. (b) denotes the measured bit
error rate obtained for intervals of 500 samples by running 10
different sequences through the simulation system. b_n is the
theoretical error rate as calculated from the measured output signal to
noise ratio under a Gaussian noise assumption. b_o is similarly the
optimum achievable error rate calculated from the theoretical
obtainable output signal to noise ratio.

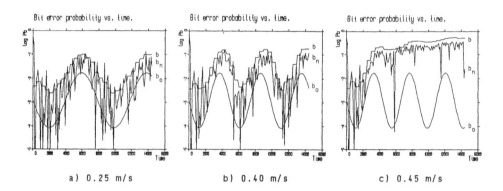

a) 0.25 m/s   b) 0.40 m/s   c) 0.45 m/s

Figure 7. Bit error probability vs. time for different velocities.

4.   CONCLUSION

A brief description of how the multipath interference problem is met by
a combination of beamforming and adaptive equalization is given. The
beamformer offers suppression of long delay multipath interference
while the equalizer counteracts remaining symbol interference due to
closely separated paths. The tracking performance of the equalizer is
addressed by simulations employing a discrete multipath model.
    For the chosen geometry the equalizer accommodates a transmitter
velocity of 0.4 m/s. Even though this limit allows for realistic ROV
motion, the tracking performance is a critical issue calling for better
equalizer algorithms.

5.   ACKNOWLEDGEMENTS

The author wishes to thank his colleagues Knut Rimstad and Arne Solstad
for cooperation throughout this work. Special thanks to Gerd U.
Kjaerland and Inghild Øien for preparing this manuscript and its
illustrations.

6.   REFERENCES

Collins, J.S., Galloway, J.L. and Balderson, M.R. (1985) 'Auto
aligning system for narrow band acoustic telemetry`, Conf. Rec.
Oceans '85, San Diego.

Haykin, S. (Ed.) (1985) 'Array Signal Processing`, Prentice-Hall
Inc., N.J., ISBN 0-13-0464482-1.

Haykin, S. (1986) 'Adaptive Filter Theory`, Prentice-Hall Inc.,
N.J., ISBN 0-13-004052-5.

Honig, M.L. and Messerschmitt, D.G. (1984) 'Adaptive Filters,
Structures, Algorithms and Applications`, Kluwer Academic
Publishers, ISBN 0-89838-163-0.

Monzingo, R.A. and Miller, T.W. (1980) 'Introduction to Adaptive
Arrays`, J. Wiley & Sons, ISBN 0-471-05744-4.

Proakis, J.G. (1983) 'Digital Communications`, McGraw-Hill,
ISBN 0-07-066490-0.

Roberts, S.J. (1983) 'An echo cancelling technique applied to an
underwater acoustic data link`, Ph.D. Thesis, Heriot-Watt
University, Edinburgh.

Sandsmark, G.H. and Solstad, A. (1987) 'Counteraction of multipath
interference by a combination of beamsteering and adaptive
equalization`, Proc. Inst. Acoustics, Vol. 9, Pt. 4.

Widrow, B. and Stearns, S.D. (1985) 'Adaptive Signal Processing`,
Prentice-Hall Inc., N.J., ISBN 0-13-004029.

UNDERWATER ACOUSTICS FOR SUBMERSIBLES

ROBERT P. GILBERT[1] and YONGZHI XU[1]
DAVID H. WOOD[2]

ABSTRACT. Sound is the only known form of energy that can be
efficiently transmitted through water. It is the main tool used for
underwater investigations whether they be probing the structure of the
bottom, charting the bottom, or searching for sunken vessels.
Acoustical methods for the tracking and identification of submersibles
and transmitting distress signals are of great interest to those using
submersibles. The increased use of submersibles for commercial and
recreational purposes may create new and hazardous situations.
Examples of such activities include an increase in scuba diving,
underwater exploration, and the introduction of submarine cargo ships
or oil tankers. Measures must be considered for locating hazards and
avoiding accidents. To this end, a better understanding of how sound
propagates in the ocean is not only useful but actually necessary.

1.   INTRODUCTION

The attenuation of sound is an order of magnitude smaller in water than
in air. Primarily for this reason acoustic energy is propagated rather
efficiently through water. Indeed, sound is the only known form of
energy that can be efficiently transmitted through water. This
property of sound has made it a rather attractive means for underwater
investigations whether they be seismic, charting or searching. The use
of acoustical methods for tracking and identification of submersibles
and transmitting distress signals is of great interest to those using
submersibles. Furthermore, an increase is expected in many underwater
activities for private and federal purposes. The increased use of
submersibles for commercial and recreational purposes may create new
and hazardous situations. Examples of such activities include an
increase in scuba diving, underwater oil exploration and the

[1] Dept. of Mathematics, University of Delaware, Newark, Delaware, USA.

[2] Dept. of Computer & Information Sciences, University of Delaware,
    Newark, Delaware, USA   and   Code 3122, New London Laboratory, Naval
    Underwater Systems Center, New London, Connecticut, USA.

D. A. Ardus and M. A. Champ (eds.), Ocean Resources, Vol. II, 135–146.
© 1990 *Kluwer Academic Publishers. Printed in the Netherlands.*

introduction of submarine cargo ships or oil tankers to bring oil from
Alaskan oil fields under the polar ice cap. Moreover, the storing of
oil and other expendables in underwater containers or structures and
the possibilities for collisions with submarine cargo ships and trawls
mandates that new precautions must be considered for locating hazards
and avoiding accidents. To this end, a better understanding of how
sound propagates in the ocean is not only useful but actually
necessary.

The submersible identification problem falls into the general area
of problems known as inverse problems. Another such problem concerns
the determination of ocean characteristics by measuring their effects
on waves propagating from a known source. For example, the
determination of the index of refraction by measuring the amplitude of
sound emitted by a fixed transducer at a known distance is another
example of an inverse problem. New analytical solutions obtained by
our transmutation methods (Gilbert and Wood, 1986; Gilbert, Wood and
Xu, 1988; Gilbert and Xu, 1987, 1988, 1988 (preprint)) may be used to
generate algorithms for solving these problems.

2.   DIRECT AND INVERSE SOUND FIELD PROBLEMS IN A FINITE DEPTH OCEAN

For this Conference, we report on our efforts to compute
representations of sound fields in an ocean of finite depth using the
MATLAB (PC-MATLAB, 1986) programming language on personal computers.
These tools apply to both the problem of computing the sound field in
the ocean from knowing its source and scatterers, and the problem of
infering the source and scatterers from measurements of the sound
field.

In the theoretical study of the sound field produced by a source
in an ocean, the problem of a point source in an ocean of constant
depth has been investigated very thoroughly (cf. Ahluwalia and Kellar,
1977; Brekhovskiskh, 1980). Probably the reason for this is that it is
a good approximation for any sound source far away in the real ocean.
In the past, this has been reasonable; however, it is anticipated that
in the future it will be increasingly important to deal with quieter,
dispersed, sources of sound in the ocean. This means that one must be
closer to the sources in order to be able to detect them.

Before, observations were made from so far away that one could
assume that the sound source was simply a point source. Now one must
assess the effects of distributed sources of sound, that is, one must
consider the shapes of the sound sources and scatterers. This
motivates investigation of two kinds of problems:

1.   The direct propagation problem, i.e. if we know the shape of an
      object, what is the sound field produced or scattered by it? What
      can we expect to detect from a reasonable distance? This problem
      is addressed by finding an operator T that maps the boundary of
      the obstacle and the incident field onto the sound field further
      away from the source.

2.   The inverse scattering problem, i.e. if we have sampled the
      scattered sound at a distance, can we say anything about the shape
      of the obstacle, that is, can we invert the operator T to
      reconstruct the shape of the scattering obstacle?

      For the direct problem, much investigation has been done.
However, as F. Dias pointed out (Dias, 1988): "Among the greatest
challenges faced by ocean acoustic engineers are the problems of three-
dimensional propagation of sound in the ocean and scattering of sound
waves from rough surfaces, such as the sea surface and the bottom of
the ocean." Recently, the technique of **transmutation** has been used by
the authors and others (cf. Duston et al, 1987; Gilbert and Wood, 1986;
Gilbert and Xu, 1987, 1988 preprint; Gilbert, Xu and Thejll, 1988;
Gilbert, Xu and Wood, 1988; Xu, 1988 preprint) for such problems. The
next sections of this paper demonstrate practical techniques for
numerical computation based on transmutation.
      Recently there has been a lot of progress made on the inverse
scattering problem for time-harmonic acoustic waves in the whole space
(cf. Brekhovskiskh, 1980; Colton and Kress, 1983). However, little of
this carries over to problems in an ocean of finite depth, because the
interaction of the surface and bottom of the ocean with the sound field
has a drastic effect on the far field pattern. These surfaces form a
waveguide trapping a field that mostly consists of a finite number of
"normal modes" somewhat like those of an organ pipe. The rest of the
field is evanescent, suffering exponential decay with range. The
recent work of Gilbert and Xu has dealt with the problem of extending
inverse scattering results to more realistic models of the ocean that
include waveguide effects due to finite depth.

3.   NUMERICAL TRANSMUTATION USING PERSONAL COMPUTERS

Our objective is to demonstrate practical and efficient methods for
sound field computations for modeling an ocean of finite depth with
sound speed that depends on depth. Our methods are efficient because
they exploit the fact that the transmutation kernel is independent of
range and can be computed in time proportional to the square of the
depth of the ocean. Our methods are practical because they make modest
demands for computing power: they are done on personal computers.
      The transmutation technique we use transforms solutions of
equations with constant coefficients into solutions of equations with
variable coefficients. It is well know that, for example, Fourier
transforms and Hankel transforms can map solutions of one equation into
solutions of a different equation. Transmutation is a generalization
of this, where in our context, we custom make a transformation that
maps the normal modes of an ocean model with constant sound speed
(these are certain cosine functions) into the normal modes of an ocean
model with a sound speed that varies with depth, this sound speed being
provided from measurements or estimates at the time and place of
computation.
      At first transmutation sounds too good to be possible, but we
quickly see that it requires solving a partial differential equation.

However, it turns out that it is not very difficult to apply the
numerical method of finite differences to solve this equation. Once
this result is obtained, it can be used and reused for computations
regardless of the size of the horizontal range.

In order to consider realistic oceans with an index of refraction
which is depth dependent we must solve the depth-dependent Helmholtz
equation

$$\Delta u + k^2 n^2(z)u = 0$$

which governs the sound field in the ocean. Our solution may be
generated by means of the transmutation

$$\phi(z) = \psi(z) + \int_{z=z_b}^{z} K(z,s)\phi(s)\,ds$$

where the transmutation kernel $K(z,s)$ satisfies the Gelfand-Levitan
equation

$$\frac{\partial^2 K}{\partial z^2} + \frac{\partial^2 K}{\partial s^2} + k^2[n^2(z)-1]\,K = 0$$

and the characteristic conditions

$$2\frac{\partial}{\partial z}K(z,z) + k^2[n^2(z)-1] = 0$$

and

$$2\frac{\partial}{\partial z}K(z,-z+2h) + k^2[n^2(z)-1] = 0$$

We apply the method of finite differences in the MATLAB function
TRFDHR to find the kernel $K(z,s)$ for any index or refraction $n(z)$ that
the user may provide to model his local situation. Full details of our
MATLAB functions and their underlying methods can be found in (Gilbert
et al, in preparation).

In order to obtain the propagating field that satisfies the
Helmholtz equation, we compute the modal expansion of the sound field

$$i\pi \sum_{n=0}^{N} \phi_n(z)\phi_n(z_o)H_m^1(k\alpha_n r)$$

where the $\phi_n(z)$ are the normal modes (eigenfunctions of the separated
z-equation) for the variable index of refraction $n(z)$, and the α_n are
the corresponding eigenvalues. The normal modes may be computed by
transmutation (Gilbert and Wood, 1986), and we have developed a MATLAB

program which does exactly this. The $\phi_n(z)$ are the solutions of the
depth-variable equation

$$\phi_{zz} + k^2 n^2(z)\phi = \alpha_n \phi$$

and the boundary conditions

$$\phi(0) = 0, \ \phi_z(z_b) = 0$$

where the α_n are the corresponding eigenvalues.

To find the normal modes $\phi_n(z)$ for our given index of refraction
$n(z)$ we apply transmutation to functions that satisfy the above
equation when the index of refraction $n(z)$ is a constant, namely

$$\psi(z,\alpha) = \cos(\alpha(z_b - z))$$

The result of the transmutation is

$$\phi(z,\alpha) = \psi(z,\alpha) + \int_{z=z_b}^{z} K(z,s)\psi(s,\alpha)ds$$

It can easily be seen (Gilbert and Wood, 1986) that $\phi(z,\alpha)$ always
satisfies the boundary condition $\phi_z(z_b,\alpha) = 0$ for all α, and we
interpolate among samples to find values of α_n such that the first
boundary condition is also satisfied, namely $\phi(0,\alpha_n) = 0$. These values
of α_n are the eigenvalues corresponding to $n(z)$, and the functions
$\phi(z,\alpha_n)$ are the normal modes (after they have been L_2 normalized).

When the normal modes and eigenvalues are known, the sound field is
computed using the MATLAB function (Gilbert et al, in preparation) HZMO
to obtain

$$i\pi \sum_{n=0}^{N} \phi_n(z)\phi_n(z_o)H_m^{(1)}(\alpha_n r)$$

4.   OCEAN ACOUSTIC COMPUTATIONAL EXAMPLES

We now present computations that were done using our MATLAB functions.
First of all, our MATLAB function TRFHDR was used to approximate the
transmutation kernel in the case when $n(z)$ is a constant and the
results were compared to those from our MATLAB function TRANHRCO based
on the exact analytic representation of the kernel for this case
(Gilbert and Wood, 1986; Eqn. 30). These results are shown in figures

1 and 2, which show that the error of using finite differences decreases as the square of the mesh size.

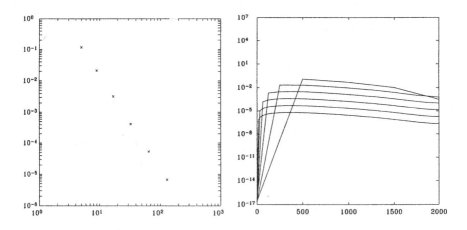

Figure 1.   Figure 2.

Figure 1. Relative error in computing the transmutation kernel $K(z,s)$ by finite differences. The vertical axis measures the ratio of the maximum of the difference of the approximate and exact kernels divided by the norm of the exact kernel. The horizontal axis shows the number of divisions of depth that were used in approximating the kernel.

Figure 2. Relative error in computing the transmutation kernel $K(z,s)$ by finite differences is shown for $z=0$, which is important for computing the eigenvalues. The vertical axis measures the error, the horizontal axis represents the variable s. The different curves are from dividing the depth into 4, 8, 16, 32, 64 and 128 parts.

      Computations will now be presented for two idealized ocean examples and compared to computations done using SNAP (Jensen and Ferle, 1979), a general purpose ocean normal mode computer model that was run on a VAX computer. Eigenvalue computations for these two idealized oceans were featured in (Robinson and Wood, 1988). The ocean sound speeds (proportional to the reciprocal of the index of refraction) for the two examples are shown in figures 3 and 4, taken from (Robinson and Wood, 1988).

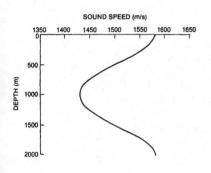
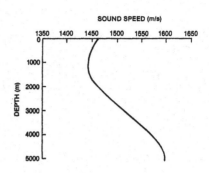

Figure 3.   Figure 4.

Figure 3. Sound speed as a function of depth for idealized symmetric
sound channel used for example 1.

Figure 4. Sound speed as a function of depth for idealized deep ocean
channel used for example 2.

    The normal modes computed by our MATLAB function (Gilbert et al,
in preparation) MOTRHR are compared to those from SNAP in figures 5-8.

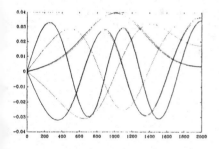
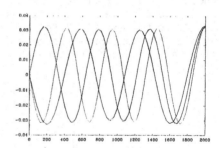

Figure 5.   Figure 6.

Figure 5. Comparison of normal modes 1-5 for example 1 as computed by
MATLAB functions and by SNAP, a large general purpose normal mode
computer model.

Figure 6. Comparison of normal modes 6-8 for example 1 as computed by
MATLAB functions and by SNAP, a large general purpose normal mode
computer model.

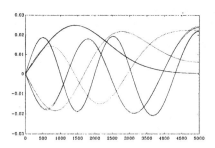 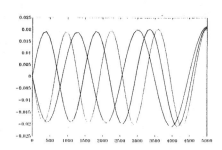

Figure 7.   Figure 8.

Figure 7. Comparison of normal modes 1-5 for example 2 as computed by
MATLAB functions and by SNAP, a large general purpose normal mode
computer model.

Figure 8. Comparison of normal modes 6-8 for example 2 as computed by
MATLAB functions and by SNAP, a large general purpose normal mode
computer model.

     The normal modes are, of course, only an intermediate result, but
since they are independent of range, this allows us to then compute the
sound field wherever we like. Figures 9 and 10 show the intensity of
the sound field for example 1 for the sound source and receiver at the
same depth as the horizontal range varies from 0 to 10 km. In this
example we take the sound to consist of a single frequency of 3 Hz.

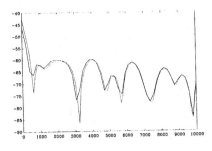 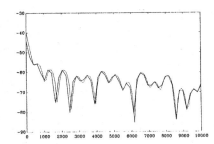

Figure 9.   Figure 10.

Figure 9.   Comparison of sound intensity on a logarithmic scale for
example 1 as computed by MATLAB functions and by SNAP, a large general
purpose normal mode computer model.   In this figure, the source and
receiver are both at a depth of 100 m.

Figure 10.   Comparison of sound intensity on a logarithmic scale for
example 1 as computed by MATLAB functions and by SNAP, a large general
purpose normal mode computer model.   In this figure, the source and
receiver are both at a depth of 1000 m.

   Similarly, figures 11-14 show intensity versus range for four
different depths for example 2.   In this example we model a sound
frequency of 1.2 Hz.

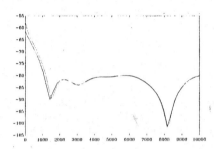
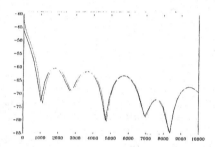

Figure 11.   Figure 12.

Figure 11.   Comparison of sound intensity on a logarithmic scale for
example 2 as computed by MATLAB functions and by SNAP, a large general
purpose normal mode computer model.   In this figure, the source and
receiver are both at a depth of 100 m.

Figure 12.   Comparison of sound intensity on a logarithmic scale for
example 2 as computed by MATLAB functions and by SNAP, a large general
purpose normal mode computer model.   In this figure, the source and
receiver are both at a depth of 1000 m.

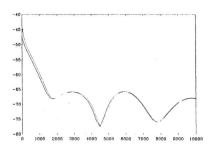 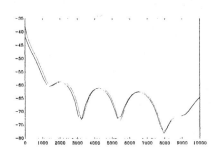

Figure 13.   Figure 14.

Figure 13. Comparison of sound intensity on a logarithmic scale for
example 2 as computed by MATLAB functions and by SNAP, a large general
purpose normal mode computer model. In this figure, the source and
receiver are both at a depth of 2500 m.

Figure 12. Comparison of sound intensity on a logarithmic scale for
example 2 as computed by MATLAB functions and by SNAP, a large general
purpose normal mode computer model. In this figure, the source and
receiver are both at a depth of 5000 m.

5.   ACKNOWLEDGEMENT

The first two authors' research was supported in part by Sea Grant
NA86AA-D-SG040 and the third author's research by NUSC IR/IED Project
A65045.

6.   REFERENCES

Ahluwalia, D. and Keller, J. (1977) 'Exact and asymptotic
    representations of the sound field in a stratified ocean', Wave
    Propagation and Underwater Acoustics, Lecture Notes in Physics
    70, Springer.

Angell, T.S., Colton, D. and Kress, R. (1988) 'Far field patterns
    and inverse scattering problems for imperfectly conducting
    obstacles', (preprint).

Brekhovskiskh, L. (1980) 'Waves in Layered Media`, 2nd Edition, Academic Press, New York.

Colton, D. and Kress, R. (1983) 'Integral Equation Methods in Scattering Theory`, John Wiley, New York.

Colton, D. and Monk, P. (1985) 'A novel method of solving the inverse scattering problem for time-harmonic acoustic waves in the resonance region`, SIAM J. Appl. Math., 45, pp.1039-1053.

Colton, D. and Monk, P. (1985) 'A novel method of solving the inverse scattering problem for time-harmonic acoustic waves in the resonance region: II`, SIAM J. Appl. Math., 46, pp.506-523.

Dias, F. (1988) 'Mathematics of general ocean engineering`, SIAM News.

Duston, M.D., Gilbert, R.P., Verma, G.H. and Wood, D.H. (1987) 'Direct generation of normal modes by transmution theory`, Computational Acoustics: Algorithms and Applications, Vol. 2, pp.389-402.

Gilbert, R.P. and Wood, D.H. (1986) 'A transmutational approach to underwater sound propagation`, Wave Motion 8, pp.383-397.

Gilbert, R.P. and Xu, Y. (1987) 'Starting fields and far fields in ocean acoustics`, Wave Motion 8, (to appear).

Gilbert, R.P. and Xu, Y. (1988) 'Dense sets and the projection theorem for acoustic harmonic waves in homogeneous finite depth oceans`, (preprint).

Gilbert, R.P. and Xu, Y. (1988) 'The propagation problem and farfield patterns in a stratified finite depth ocean`, (preprint).

Gilbert, R.P., Xu, Y, and Thejll, P. (1988) 'An approximation scheme for three-dimensional scattered wave and its propagating far-field pattern in a finite depth ocean`.

Gilbert, R.P., Xu, Y, and Wood, D.H. (1988) 'Construction of approximations to acoustic Green's functions for nonhomogeneous oceans using transmutation`, Wave Motion 10, pp.285-297.

Gilbert, R.P., Wood, D.H. and Xu, Y. 'Underwater Acoustics normal mode modeling using transmutation and PC-MATLAB` Sea Grant Report, Univ. of Delaware, (in preparation).

Jensen, F.B. and Ferle, M.C. (1979) 'SNAP: The SACLANTCEN normal-mode acoustic propagation model`, SACLANTCEN Memorandum SM-121, SACLANT ASW Research Center, La Spezia, Italy.

Robinson, E.R. and Wood, D.H. (1988) ´Generating starting fields
  for parabolic equations`, J. Acoust. Soc. Amer. 84, pp.1794-
  1801.

Xu, Y (1988) ´The propagating solution and far field patterns
  for acoustic harmonic waves in a finite depth ocean` (preprint).

Xu, Y (1988) ´An injective far field pattern operator and inverse
  scattering problem in a finite depth ocean` (preprint).

´PC-MATLAB User´s Guide` (1986) Mathworks, South Ntick, MA.

PART III

Subsea Automation Technology

UNDERWATER OPTICAL SURVEYING AND MAPPING - TRANSFERRED RESEARCH EFFORT
STRENGTHENS OCEAN TECHNOLOGY DEVELOPMENT

J. LEXANDER, O. STEINVALL, S. SVENSSON, T. CLAESSON,
C. EKSTROM AND B. ERICSSON
The Swedish Defence Research Establishment
Linkoping
Sweden

1. INTRODUCTION

The following transferred research effort referred to is a 10-year co-
operative agreement between the Swedish Defence Research Establishment
(FOA) and the civilian co-partner, the National Board of Technical
Development (STU). The latter is the largest funding organization in
Sweden for the support of technical R&D.
        The primary aim of our co-operative agreement has been to promote
a transfer of FOA's military competence in underwater technology to
ocean technology industries. The competence areas involved were
mainly:

-   Hydro-acoustics
-   Hydro-optics
-   Remote control systems
-   Energy sources
-   Diving physiology and techniques.

        The content of the agreement consisted of a number of well defined
projects. When these were finished new specific R&D-projects were
chosen. A large part of the work concerned marine survey technology.
Two of the larger projects, one in the field of seismology and one in
diving physiology were run during the whole 10-year period. The diving
project concerned the possible use of hydrox (mixture of hydrogen and
oxygen) as a breathing gas for deep diving.
        Prototypes and ideas have been transferred to Swedish industries.
Co-operative work for exchanging knowledge and development results have
been carried out with Norwegian ocean technology companies and
institutions. The R&D carried out has, to a large extent, been
directed towards the oil and gas offshore industry.
        The projects we would specifically like to outline in this paper
are:

-   Airborne laser system for bathymetric and bottom topographic
    mapping.

D. A. Ardus and M. A. Champ (eds.), Ocean Resources, Vol. II, 149–154.

–   Instrument designed for multiple optical water quality
    measurements.

    Optical water quality measurement is a reference which may be
combined with the evaluation of the technical performances of laser
bathymetry and other hydro-optical applications as well as water
quality control for environmental protection.
    A study of laser bathymetric techniques has been carried out
together with the Continental Shelf Institute in Trondheim, Norway
(Steinvall and Klepsvik, 1984).

2.   AIRBORNE SCANNING LASER HYDROGRAPHIC SURVEYING

The system concept is illustrated in the block diagram in figure 1.

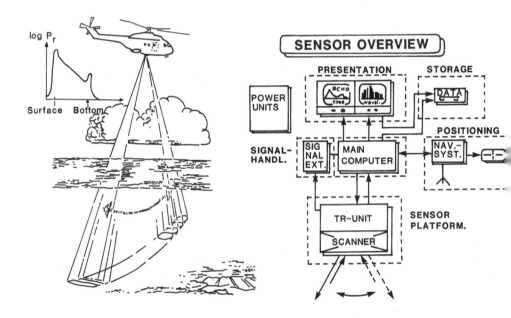

Figure 1.   System overview for a laser bathymeter for depth charting.

    The principle is simple: a short laser pulse, with emission in the
green wavelength, 532 nanometer (nm), is transmitted towards the water
surface and penetrates it to reach the sea bottom. The laser light
reflections from the water and bottom surfaces are detected by the
receiver in the aircraft. The time difference measured between the two
reflections is used to calculate the water depth. The scanning of the
pulsed laser beam makes it possible to cover a broad swath which with a
flight altitude of 500 m can reach about 300 m.
    The emission wavelength lies close to the transmission maximum for
coastal waters. However, the penetration capability depends to a large

degree on the local optical water quality conditions. For coastal
waters these are to a large extent determined by the turbidity or the
particle scattering conditions. The laser bathymeter technology of
today has a normal depth penetration capacity in ordinary coastal
waters of 15-30 m. To learn more about the possible depth capacity we
need more knowledge about the optical water characteristics in
different coastal water areas extended over seasons.

3. WHY LASER BATHYMETRY?

Sweden was one of the countries, as well as, for example, USA, Canada,
Australia, which started investigations into underwater laser
reconnaissance and laser depth measurements at the end of the 1960's
and early 1970's. If we limit our interest to the depth and
topographic mapping applications the foremost advantage of the airborne
laser scanning bathymetry is the larger area coverage rate compared to
the traditional echo sounding techniques carried out from ships. The
area coverage rate is given by the laser pulse frequency and the
sounding density. With one sounding per 25 m^2 the coverage rate will
range from about 20 km^2/h with a 200 Hz laser to more than 200 km^2/h
for future high-prf lasers with 2000 Hz. The coverage rate capacity
saves substantial cost and manpower. Estimates based on studies of
laser surveyable coastal areas indicate a cost-benefit ratio of about
6:1 when comparing the survey cost per km^2 with the present echo
sounding techniques. Other advantages or points of interest that can
be noted are:

- The laser system is especially useful in narrow channels and areas
  with many small islands and skerries, which are very time
  consuming for ship survey. The laser can also in principle
  measure depths all the way up to land, which is of value for the
  small boats used for recreation. This problem is normally not
  considered with the present survey methods.

- The high speed capability of an airborne system will make it
  suitable for presurveys, which in turn will help in optimizing the
  use of all surveying systems.

- The high density of soundings improves safety by increasing the
  probability of detecting hazards to navigation.

- The increased number of soundings per unit area and the earlier
  use of this data due to increased production rates will improve
  safety. As a result the present survey fleet can spend more time
  in high priority areas like ship routes, etc.

- It is concluded that laser hydrography has about the same weather
  constraints as sonar, and can operate day or night and is only
  limited by strong winds (>20 knots), fog and high precipitation.

- An area covering sensor like a laser bathymeter will gather a very
  large data volume per mission which must be processed in the field
  to accomplish a comparable rate between collection and processing.
  It will also give the hydrographer in charge the possibility of
  optimizing flight hours during the mission. The high degree of
  processing automation and large number of soundings will support
  the development of automated chart production. Such a function
  will rely on a digital terrain model.

4.   SYSTEM PERFORMANCE - OPERATIONAL ASPECTS

The areas being charted by the laser bathymetry system have to be
selected in advance. From pre-investigations the areas are evaluated
suitable for optical surveying in accordance with the system
performances. It is important to assess the penetration depth
performance related to the key optical water quality parameters. Based
on the attenuation coefficient (c) for a collimated light beam or the
reciprocal of that value, (1/c) (called the attenuation length), one
can calculate the following penetration depths:

10-12 m, for 1/c = 0.5 m	The higher values of
	the penetration depths
16-20 m, for 1/c = 1 m	are for night operations.
27-32 m, for 1/c = 2 m	
30-35 m, for 1/c = 3 m	

The attenuation coefficient c should be measured at some points
within these areas during the time of the year the mission is planned
for.

Also included in premission planning, at least during the
preoperative work, are verification soundings by establishing some well
measured depths within the area, which can be used for calibration.
Another important preparation is to establish a positioning fixing
network based on microwave, laser or radio positioning systems. Which
system to be chosen has to be decided with regard to the base line
distances. Future operational systems will use the satellite based GPS
(Global Positioning System) for the positioning of the aircraft.

A 4-5 man group will be responsible for the short-term planning
and execution of the survey. The chief of the survey group will be an
experienced hydrographer and he will be responsible for the final
planning of "go" or "no-go" over a selected area. This decision will
be mainly based upon short term ($\leq$ 1 week) weather forecasts. He will
also, together with the pilot, be responsible for the same decision on
a daily basis. The survey pattern will be laid out in advance and the
flight lines will depend on the type of mission. For a basic survey
the flight will be parallel to the depth contours to maximize their
length. For reconnaissance surveys, the lines will be perpendicular to
the depth contours if the purpose is cross-checking historical data.

When looking for shoals and hazardous areas the survey speed should be low to ensure an increased sounding density.

5.   OPTICAL WATER QUALITY MEASUREMENTS

Every field experiment with optical underwater systems, carried out at FOA, has been simultaneously followed up by in situ measurements of at least the attenuation coefficient (c) of a collimated light beam. When required, the diffuse daylight attenuation (K) has been measured. That gives the possibility of comparing system performances achieved at different occasions and localities and explaining different outcomes. In testing the laser bathymetry system the detected laser light response from the vertical water column versus time is itself a kind of description of the optical water attenuation characteristics. However, this response is very much connected to the optical features of the specific laser bathymetry system used. To generalize such a recorded attenuation to the usually used attenuation key factors might be difficult. So laser bathymeters will not be a substitute for the regularly used methods for optical water quality measurements.

To facilitate the reference measurements FOA has developed a system which alternatively can be used from a ship or a helicopter. It has a measurement depth capacity of 100 m. The measurement time, when continuously moving the instrument down to this depth and up again, is less than 8 minutes.

The instrument is called HOSS which stands for the Hydro-Optical Sensor System. It measures simultaneously transmission, forward scattering, back scattering and the irradiance of the downwelling light. Temperature and salinity measurements are added to be correlated with the optical ones. The display of the calculated optical values in real-time might show the variations of the attenuation coefficient of a collimated light beam (c), the scattering coefficient (b) and the attenuation coefficient of the diffuse light (K), all versus depth. The water attenuation factor is most nearly correlated to the depth performance of the laser bathymeter.

In addition to the regular use of the HOSS for reference measurements it will be used to map variations in optical water quality round the Swedish coastal areas as a function of season, weather and other parameters.

6.   CONCLUSION

The use of optical techniques for underwater and bottom observations, inspections and mapping will increase. A cause of that trend is that more efficient optical light sources and sensors have become available and are in practical use. Also we have a better knowledge about how to adapt the use of optical underwater devices based on propagation constraints. Furthermore the possibilities of signal processing have developed dramatically. It is now possible to analyze and store thousands of high speed returns per second in real time in airborne

equipment. We are also in a better situation to combine hydro-
acoustics and hydro-optics to result in an improvement in the total
efficiency.
     This paper has informed primarily about the capacity of a laser
bathymetry system and about optical measurements for the planning of
its use. It also informs about the possibilities for a small country
to make developments in expensive technology areas and strengthen this
activity through a competence transfer from a larger activity and
market area, in this case from the military area.

7.   REFERENCE

     Steinvall, O., Klepsvik, J.O. et al (1984) 'Advanced technology
          for hydrographic mapping', FOA Report C 30371-E1, ISSN 0347-
          3708, National Defence Research Institute, Sweden, Continental
          Shelf Institute, Norway.

AUTOMATED PROCESSING AND INTERPRETATION OF SENSORY DATA FOR DEEP OCEAN
RESOURCE DEVELOPMENT

G.T. RUSSELL, G.A. SHIPPEY, D.M. LANE, L.M. LINNETT, and
A.J. RICHARDSON
Heriot-Watt University
Department of Electrical & Electronic Engineering
Edinburgh, EH1 2HT, Scotland, UK

ABSTRACT. Ongoing work is described regarding the use of algorithmic
and knowledge based system (KBS) techniques for the processing and
interpretation of data derived from sensors measuring the water column,
sea bed and sub-bottom ocean environments. This work comes from part
of a major initiative spanning fundamental, strategic and applied
research programmes in sub-sea automation and advanced robotics for
sub-sea deployment. The particular aspect of the research covered in
this paper relates to the digital processing of sector and sidescan
sonar and high-resolution sub-bottom profiler recordings made for
survey purposes, together with the knowledge-based interpretation of
these acoustic images. The role of a knowledge based system in
automated data interpretation is described, along with one knowledge
based system architecture (the blackboard system) which has
successfully been used when processing sector scan sonar images.

1.   INTRODUCTION

The rich and diverse resources which are contained within the ocean
environment surrounding our shores have resulted in mankind involving
himself in many and various forms of direct intervention activity.
Typically the motivation for this relates to food production (fishing,
farming), transport (civil and military), energy survey and production
(oil and gas, OTEC, wind/waves/tide), mineral mining (manganese
nodules), science (marine biology, geology, particle physics), dumping
(nuclear and chemical waste), salvage (archeological artefacts, pre-
nuclear steel) or recreation (sailing, diving etc.). The nature of the
intervention may be manned (e.g. divers, manned vehicles), or unmanned
(e.g. remotely operated vehicles). For most of these applications
there is a need to identify the extent of the resource prior to any
subsequent intervention and development. Carrying this out effectively
can limit the subsequent cost expressed as a function of time, expense
and ecological impact. This is particularly true for the deep ocean,
where intervention becomes significantly more problematic. There is
therefore a requirement to satisfactorily measure, pre-process and

D. A. Ardus and M. A. Champ (eds.), Ocean Resources, Vol. II, 155–169.

interpret data derived from one or a number of sensors investigating
the ocean regions of interest.

Examples of these regions need not be limited to the water column
but may also include surface, sea bed and sub-bottom areas and volumes.
During the initial reconnaissance phase such measurements are most
easily carried out from a surface vessel using long range sensors such
as low frequency sonar (for sea bed and sub-bottom), proton
magnetometer (for sea bed) and synthetic aperture radar (for ocean
surface). Subsequent more detailed measurements can be made by
deploying some form of unmanned submersible (towed, tethered or
autonomous) carrying higher definition short range sensors such as
video or high frequency sonar.

Prior to being used, the raw data derived from these sensors is
usually subject to some form of automated processing and/or
interpretation. The need for this comes about because:

(a)   the data has been corrupted by the transmission medium, spurious
      returns, or the anomalous behaviour of the measurement device
      (e.g. detection in receiver beam sidelobes, multipath reflections,
      deconvolution of received echoes with source wavelet etc).

(b)   large quantities of data have been generated. This may have to be
      transmitted over a limited bandwidth communication channel in a
      timeframe which is shorter than the channel bit rate will allow
      (e.g. acoustic communication of video information for an
      autonomous vehicle).

(c)   on site interpretation of sensory data is required for decision
      making about intervention activities in the immediate future. The
      human expertise to carry out this task may not be immediately
      available, or may take too long to apply.

Effective EEZ resource identification and development activities
can therefore require powerful tools for signal processing and
interpretation. Often it is desirable that these tools be available on
site and on line so that results can quickly be generated to allow more
rapid decision making regarding subsequent measurements. Because of
the computational cost involved many such remote sensing tasks have
resorted to guesswork backed up by subsequent post-processing carried
out at a centrally located batch processing facility.

This paper describes ongoing work which is attempting to realize
some of these signal processing and interpretation techniques in such a
way that they will be available on site and on line at the worksite,
primarily for different types of active sonar device. The techniques
encompass signal and image processing tools for received echo
deconvolution (seismic inversion), multiple echo removal and texture
analysis, in addition to novel knowledge based (KB) techniques for
automated selection of appropriate tools and processing parameters, for
detection of anomalous echoes, and for interpretation of regions of the
acoustic image within the context of the mode of operation.

In section 2, digital processing of sidescan sonar and high-
resolution sub-bottom profiler recordings is described briefly in the

context of recent results by other workers in the field and related to
the comparison work on expert systems. Section 3 describes the role
that knowledge based system techniques can play in carrying out
automated interpretation of such sonar data. Section 4 concludes with
a summary and references.

2.   COMBINING SIGNAL AND IMAGE PROCESSING TECHNIQUES FOR ACOUSTIC
     RECORD INTERPRETATION

2.1 Introduction

Until quite recently, sidescan and subbottom sonar have been poor
relations so far as digital processing of the data was concerned. The
typical output from both was a pen chart record with severely limited
dynamic range. Electronic processing was confined to slant range
correction (sidescan sonar) and the provision of "tweak-factors" to get
the most out of the paper record, within its resolution limits.
     The aim of our own work was to use computer based analytic
techniques to clarify geological structures to the interpreter, and
provide some numerical guidance as to the material encountered. We
were convinced that there exists a wealth of information in the
acoustic return which does not appear at all on the paper record.
Computer processing has been a major activity in deep seismic
interpretation and there has been successful processing of the GLORIA
survey data (Chavez, 1986) by the IOS/USGS combination. The challenge
was to see if useful results could be obtained with more modest
computing resources, on single channel data, obtained at quite
different source frequencies. The approach was to investigate the
combination of image processing techniques familiar from pattern
recognition applications with more conventional signal processing
encountered in seismic processing packages.

2.2 Research Objectives

Our research objectives were initially defined as separate tasks, each
clearly associated with a known problem in handling acoustic records.
It soon became clear that a further research objective was to clarify
the inter-relationship between the processing required for the various
tasks.

Sidescan

(a) pre-processing, TVG and correction for trace-to-trace variability.
(b) spectral and textural discrimination between "seabed types".
(c) image enhancement to emphasise linear features e.g. rock outcrops.

Subbottom Processing

(a) Pre-processing, particularly trace alignment.
(b) Spectral measurement of wavelet attenuation with frequency and
        depth.

(c) Removal of long period multiple associated with surface-tow
     vehicles.
(d) deconvolution for source wavelet shape.
(e) image enhancement to emphasise or follow geological horizons.

     The spectral/textural analysis was stimulated by the work of Pace
and colleagues at Bath University, in surface sediment classification
from sidescan echoes (Reut et al, 1985).

2.3 Methodology

Most of the acoustic recordings used for training and test data for the
research were supplied either by Ferranti ORE (Great Yarmouth and
Falmouth, Mass.) or by Dr. Larry Mayer (Dalhousie University and Rhode
Island University). The sidescan data was obtained from an FORE
100 kHz. The subbottom recordings were obtained using the FORE
Geopulse source and a separate non-directional hydrophone.
     The computer system selected for the research was the Swedish
Contextvision GOP-300 Image Processor, originally developed by Granlund
et al at Linkoping University. This machine is widely employed by
remote sensing institutions in Europe, and has a number of relevant
hardware and software features, including a powerful convolution
processor, image processing software for structural and textural
analysis, classification software, and a comprehensive colour display
capability. Individual modules have been ported to a multi-processor
VME system, since this is nearer to the target system which will be
deployed.
     A simulation model has been developed by British Geological Survey
(Dobinson et al, 1988) to provide test images for the subbottom
algorithms. This model can simulate realistic geology, together with
the acoustic signal modifications under investigation - spectral
attenuation, jitter, noise and sea-surface multiples. This model has
proved invaluable in validating (and sometimes invalidating!) the trial
algorithms.

2.4 Summary of Results

In this section an indication of the current status of each task is
given and success is illustrated in monochrome where appropriate. GOP
colour images are unfortunately not readily reproduced.

Sidescan Preprocessing - This work has partly been reported (Eaves-
Walton and Shippey, 1987). A successful TVG algorithm was developed
which made use of the intensity histogram at each range to equalise
"background" intensity.

     The most significant problem addressed was the occasional "bad
line" or bad section of an echo, where the statistics differ markedly
from that of adjacent echoes. This may be due to interference fading,
but the mechanism is not well understood. The effective solution was
the use of a statistical replacement algorithm which replaces an echo

section by a local average if its statistics are markedly different
from that of the local neighbourhood.

Sidescan Spectral & Textural Analysis - The sidescan data available was
unsuitable for spectral analysis in the Bath manner. However textural
analysis was employed to see whether the system could perform at least
as well as a human observer in separating visibly distinct regions.
The first attempt used spatial frequency and orientation was reported
(Pelag et al, 1984). Considerably improved segmentation was then
obtained using a fractal-based technique (**Fig. 1**).

Sidescan Enhancement - Good results have been obtained for sidescan
scene enhancement using standard GOP image enhancement software. Other
geological structures of more interest could be clarified in the same
manner.

Subbottom Processing-Pre-Processing - An important step in trace-to-
trace alignment. Accurate bottom tracking is carried out using a
mixture of correlation and edge detection techniques. Figure 2 shows a
section of a raw digitized recording from the Emerald Basin area in
Canada. Considerable jitter is obvious. Figure 3 shows the same
section after automatic trace alignment.

Subbottom Spectral Analysis - According to some geological models
(Dodds, 1986) spectral attenuation expressed in dB/metre is fairly
linear over a wide frequency range, with the slope dependent on the
type of sediment through which the acoustic energy passes (Schlock et
al, 1986) (Dodds, 1986). A characteristic of the sediment is therefore
attenuation slope expressed in dB/kHz/metre. The attenuation program
which exploits comb filter measurement obtains accurate measurements up
to 10 dB/kHz/meter on simulated data, over the frequency range 1-8 kHz.
The comparison of measurements on real data with core measurements has
still to be carried out. However the indications are that a useful,
rapid tool has been developed.

Subbottom Multiple Removal - Figures 5 and 6 show the effect of an
algorithm which removes the long sea-surface multiple from a section of
Narragansett Bay data. This algorithm makes heavy use of correlation
techniques to detect and quantify the multiple in the presence of
genuine subbottom echoes. A simple gap filter then carries out
multiple cancellation.

Deconvolution - The spectral analysis required for sediment
characterization can also be employed in Q-restoration, which is an
important step in the deconvolution process. The aim of this work is
to clarify subbottom geology while acoustic survey is being carried
out.

Subbottom Image Enhancement or Stacking - The aim is to reinforce the
trace-to-trace coherent component of the echo, and to average out the
incoherent scatter. The advantages of this step are threefold, firstly
providing an image with clearer geological structures, secondly

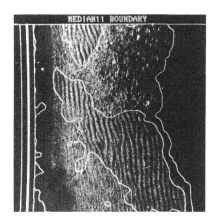

Figure 1. Sidescan image
segmented by fractal technique.

Figure 2. Raw seismic section,
enlarged around first bottom
return.

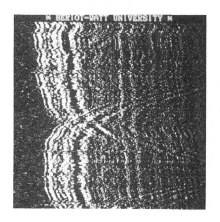

Figure 3. Section of Fig. 2
after automatic trace alignment.

Figure 4. Correlation window
tracking of second bottom
multiple.

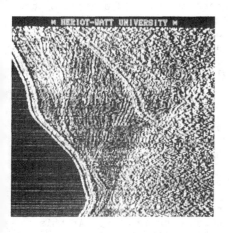

Figure 5. Narragansett Bay
section showing multiple.

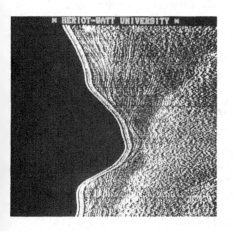

Figure 6. Narrangansett Bay
section after long multiple
cancellation.

providing a reduced dataset for the computationally intensive deconvolution operations and finally the ratio of scattered to coherent power is a potentially useful tool in geological interpretation.

Task Interdependence - The processing operations discussed in this section are not carried out in isolation and there is a proper sequence beginning with pre-processing, and going on to multiple removal, image enhancement and deconvolution. In our software these operations can be carried out in the time domain, but it is interesting how central the role of spectral measurements is to the success of the processing.

3.   KNOWLEDGE BASED INTERPRETATION OF SONAR DATA

To automate the task of sonar interpretation it is appropriate to consider the means by which learned human expertise can be incorporated with signal processing techniques such as those in Section 2. To this end techniques from Artificial Intelligence and Knowledge Based Systems (KBS) have a role to play. This section briefly discusses this role by considering the nature of human expertise for object and shadow detection in sector scan sonar images, a KBS architecture which has successfully been used for interpretation of a limited class of images (Lane, 1986) and the nature of the knowledge base employed.

3.1   The Nature of Human Expertise

The nature of the human processing activity is hierarchical, using a mix of low level scene analysis, guided by more abstract knowledge about the image, the operation of the sonar system, and the nature of the surrounding environment.
    At the lowest (and apparently subconscious) level, mechanisms of visual perception may typically identify arbitrarily shaped image regions having a mean value which is more than (i.e. object) or less than (i.e. shadow) the mean value of the surrounding area. At this level, candidate areas are constrained only in that they must possess a certain amount of connectivity in order to constitute a region. Equipped with this level of processing, the human operator can perceive candidate objects and shadows, but cannot distinguish echoes caused by bona fide objects, noise or anomalous behaviour of the sonar (e.g. echoes detected in the receiver beam sidelobes, multipath reflections or reflections of previous insonifying pulses from objects beyond the range set on the sonar). It represents a non-expert level of skill with respect to this domain, in that it can be performed by people who know nothing about sonar. It can therefore be termed perception.
    The interpretation stage, based on these perceived candidates, is a two part operation, using a conscious level of learned expert human knowledge. Initially, those regions which are objects have to be distinguished from those which are noise or anomalies. Candidate regions have to be regarded in the context of other candidate regions in the current scan, and in the context of confirmed object/shadow regions located in previous scans. Those candidates formed as a result of ambient or reverberation noise typically exhibit strong spatial

correlation around circles of constant range, and/or do not exhibit
scan to scan temporal correlation in their global axis positions.
Candidates caused by multipath anomalies have global axis positions
which move in sympathy with the motion of the sonar platform, and may
be characteristically positioned relative to other known reflectors.
Echoes caused by reflections of a previous pulse at a boundary beyond
the range setting of the sonar are similar, with mean values
significantly less than other reflections in the same region.
Similarly, sidelobe secondary scanning anomalies have characteristic
positioning relative to real reflectors. Candidates resulting from
real object reflections generally exhibit scan to scan temporal
correlation, and take on characteristic forms such as a blob or
straight line segments. Unlike the initial perception stage therefore,
this level of processing requires the use of expert knowledge specific
to the task of sonar interpretation.

     The second interpretation stage almost invariably requires the use
of information available a-priori concerning objects known to be in the
vicinity. Knowing there is a wellhead on the sea bed helps to classify
the most appropriate blob-like static candidate of appropriate
dimension and position. In the absence of such a priori information,
knowledge of object characteristics must still be employed. A moving
object cannot be a wellhead or a seabed scar, although it could be a
diver or a submersible. It is at this level that the human operator
would reason about echoes received from a diver and his bubbles say,
knowing the way in which he has been swimming, the direction of any sea
current, and therefore where the vertical column of bubbles may reside.
Such reasoning is important if the diver is to be correctly recognised
in a variety of situations. It is the need for this reasoning ability
that distinguishes interpretation from simple classification.

3.2 Blackboard Knowledge Based System Architecture

In order to express the concepts and the processing tools and
techniques required to automate this human expertise, and appropriate
knowledge based system architecture is required. This provides the
necessary programming framework. One architecture which has
successfully been employed is that of a blackboard system, which is
here described.

     The first published implementation of the blackboard architecture
was the Hearsay-I speech understanding system (Reddy et al, 1983).
Since then, a number of variants on the theme have been reported
(Erman et al, 1980) (Nii, 1986) each using a different implementation
for different applications, but with the same underlying spirit.

     The references discuss the component parts of the blackboard
approach. At the heart of the system is a global memory area, the
blackboard, employed as a hierarchical store of data structures
representing goals, data and information concerning the information
processing task. Confidence or belief attributes are usually
associated with each of these, and hence the blackboard is also a store
of facts and hypotheses (i.e certain and uncertain information). It`s
role is therefore passive.

Activity on the blackboard is achieved using independent processing modules called knowledge sources or KS for short. Each knowledge source contains expertise appropriate to one very small part of the problem domain. Thus, for sonar interpretation, a KS may contain a 2D image processing algorithm, a more conventional 1D algorithm, or some form of rule based processing. The number of knowledge sources in a knowledge base will depend on the extent of the problem domain, and the granularity of the processing knowledge employed (i.e. how much active processing is wrapped up within a single KS by the knowledge base designer).

The KS all play the role of independent experts looking at the information, data and goals stored on the blackboard. Execution is initiated by the blackboard inference mechanism. At each inference cycle, all KS examine the blackboard to see if they are able to contribute new hypotheses/facts, or provide additional confirmation or denial for those already in existence. In a single processor implementation, those KS which have something to contribute (i.e. write on the blackboard) are labelled as triggered and stored according to some priority mechanism in a list called the agenda. To execute the next KS, the inference mechanism merely activates the KS currently at the top of the agenda. Once running, a KS must complete execution without interruption, and is therefore indivisible in operation. The result of a KS execution will be the assertion of information or goals on the blackboard, which may form the trigger field for other KS. Such an assertion is termed a blackboard event. Each event may provide support for a hypothesis which is in agreement with other support, or in disagreement.

Execution continues in this way until there are no more KS waiting. The blackboard system is then said to be idle. In the absence of any further goals, data or knowledge, the competing hypotheses expressing the most confidence are taken as the best guess at the problem solution. If the knowledge base (i.e. KS) or the furnished data are insufficient then only a subset of the system goals may have been satisfied, leaving a partial solution.

Support for a hypothesis is maintained using links on the blackboard (i.e. pointers between data structures), showing which KS were responsible for its existence, and the confidence they have that the hypothesis is correct. Goals, data and information, therefore, can be thought of as a connected network of hypothesis nodes, each connection representing support provided by a KS. As KS execution proceeds, so 'islands' of hypotheses evolve, until eventually these independent fragments expand and merge to form the overall solution. In this respect the blackboard architecture employs an incremental approach to information processing, where any or all of the hypotheses and facts can be completely revised at any stage. It functions as both a means of communicating between KS, and as a store of the current state of the processing task.

3.3 Perception and Interpretation Knowledge Base

As with human expertise, a hierarchical approach is appropriate for
expressing a sonar perception and interpretation knowledge base within
a KBS framework such as the blackboard (Fig. 7).

At the lowest level, raw sonar image data is placed on the
blackboard, and operated upon by image processing knowledge sources
forming the first part of the perception operation. These KS contain
expertise in pre-processing of the sonar scan, and segmentation to
identify candidate object and shadow regions. The second stage of the
perception operation involves describing these regions in some more
abstract quantitative way, providing the basis for subsequent
interpretation operations. These feature calculating KS take the
segmented image information, and produce a series of sightings which
contain numerical measures of position, shape and pixel statistics for
the hypothesised region. The role of the KS implementing each of these
tasks in combination is therefore analogous to the subconscious, low-
level vision processing of the human operator.

The first stage of the interpretation process involves the
classification of each candidate object and shadow sighting into
object/ shadow, anomaly and noise components. These KS therefore take
the position, shape and pixel statistic values for each candidate, and
use rule based and algorithmic knowledge to implement the criteria
described in Section 2.1. The second stage of the interpretation
process takes those sightings classified as object or shadow, and
attempts to identify them at the symbol level. These KS therefore use
rule based and algorithmic knowledge to perform scan to scan
correlation of object and shadow sightings and to implement simple
reasoning functions based on the 3D world model, its expected
appearance viewed through the sonar, and the candidates that have been
identified. Such processing would make use of appropriate pattern
recognition distance metrics as a means of comparing individual feature
descriptions.

In tandem with these activities, a model driven mode of processing
uses information concerning the vehicle and sonar position, orientation
and status, and the topography of the local marine environment, to
select the most appropriate image processing tools and to calculate the
parameter values required. Low level signal processing is therefore
guided according to more abstract forms of information and knowledge.

Further description and details regarding the implementation of
such a knowledge base for the sector scan sonar application can be
found (Lane, 1986) (Russell and Lane, 1986). An alternative KBS
architecture which distributes the blackboard across a number of
processing nodes in an attempt to enhance the speed performance of the
knowledge base can be found (Lane et al, 1988).

4. CONCLUSION

This paper has described some of the processing tools and techniques
relevant to automated interpretation of active sonar data for remote
sensing of EEZ resources. Some of the techniques are on the forefront

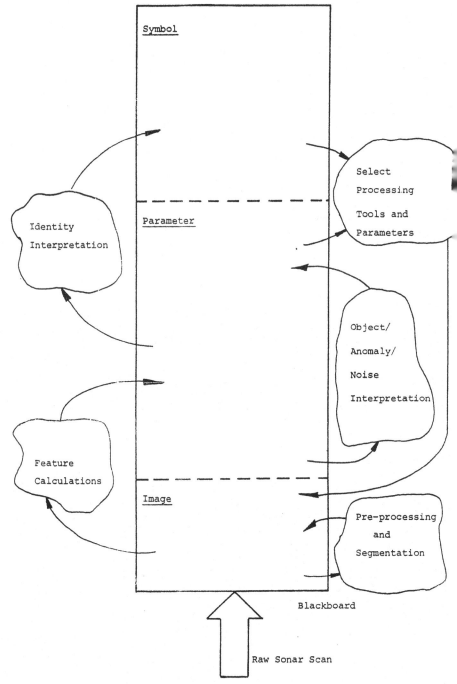

Figure 7.   Knowledge source classification for sonar perception
interpretation.

of current thinking, and will take some years to be verified and
implemented within marketable systems. Others are much closer to
practical implementation and could be expected to appear in a much
smaller timescale, dependent upon the market perceptions of its
utility.

It has been predicted that desktop computers with the power of a
Cray I supercomputer will one day be readily available. There can be
no doubt therefore that as the processing capabilities of small,
compact computer systems increases, so the feasibility of implementing
computationally costly processing techniques in portable systems
becomes more attractive. In this way, greater functionality can be
placed at the disposal of human operators working in remote situations
without the need for reliable or high bandwidth communication.
Research into techniques such as those described here is therefore
worthwhile, because it prepares the ground necessary in order to
utilise this powerful hardware, and hence seize market opportunities in
EEZ development.

5.   ACKNOWLEDGEMENTS

The research on subbottom image processing was carried out under
contract to Ferranti Ocean Research Ltd., with support from the
Offshore Supplies Office of the U.K. Dept. of Energy. FORE have kindly
given permission for publication of some of the material included in
this paper. The research was also carried out in close collaboration
with British Geological Survey.

The work on knowledge-based sonar interpretation has been funded
by the Marine Technology Directorate of the U.K. Science and
Engineering Research Council, the Offshore Supplies Office of the U.K.
Department of Energy, and a consortium of industrial partners from the
offshore energy exploration and recovery community. This is within a
Managed Programme of Research addressing many aspects of advanced subsea robotics under the title of the Automation of Sub-Sea Tasks.

The geological inspiration for the work reported here came
especially from Dr. Larry Mayer (Dalhousie University) and Dr. Alan
Dobinson (British Geological Survey). Pete Lowell of FORE, Falmouth,
was the first to formulate the research objectives for the subbottom
work. Other valuable assistance was given by Colin Graham of BGS, and
Dr. Ken Davidson, formerly in the Department, and now on the BGS staff
in Edinburgh

The authors are also indebted to the many other research
associates and colleagues for their combined efforts toward the
department marine technology programme.

6.   REFERENCES

Chavez, P.S. (1986) 'Processing techniques for digital sonar
    images from GLORIA photogrammetric engineering and remote
    sensing', Vol. 52, No. 8, pp.1138-1145.

Dobinson, A., Graham, C., Shippey, G.A. (1988) 'A seismic model
    for generating synthetic measurements using a Contextvision
    GOP300 image processor', Colloquium on simulation techniques
    applied to sonar, IEE, London, pp.5/1-5/4.

Dodds, D.J. (1986) 'Attenuation estimates for high-resolution
    subbottom profilers', Bottom Interacting Ocean Acoustics, ed
    Kuperman and Jensen, Plenum Press Marine Science Series, New
    York, Vol. 5, pp.173-191.

Eaves-Walton, C., and Shippey, G.A. (1987) 'Digital image
    processing for sidescan sonar data analysis', Proc. 5th Int.
    Conf. on Electronics for Ocean Technology, Edinburgh, pp.203-
    210.

Erman, L., Hayes-Roth, F., Lesser, V.R., Reddy, D.R., (1980) 'The
    HEARSAY II speech understanding system: integrating knowledge to
    resolve uncertainty', Computing Surveys, Vol. 12, No. 2.

Lane, D.M., (1986) 'The investigation of a knowledge based system
    architecture in the context of a subsea robotic application',
    Ph.D. thesis, Dept. Electrical & Electronic Engineering, Heriot-
    Watt University, Edinburgh.

Lane, D.M., Chantler, M.J., McFadzean, A.G., Robertson, E.W.
    (1988) 'A distributed problem solving architecture for sonar
    image interpretation', Underwater Technology, Vol, 14, No.3,
    Journal of the Society For Underwater Technology.

Love, P.L. and Simaan, M. (1985) 'Segmentation of a seismic
    section using image processing and artificial intelligence
    techniques', Pattern Recognition, Vol. IX, No. 6, pp.409-419.

Nii, H.P. (1986) 'Blackboard systems: the blackboard model of
    problem solving and the evolution of the blackboard
    architecture', The A.I. Magazine, pp.38-53.

Pelag, S. Naor, J., Avnir, D. (1984) 'Multiple resolution texture
    analysis and classification', IEEE Trans. on Pattern Analysis
    and Machine Intelligence, Vol. PAMI-6.

Reddy, D.R., Erman, L.D., Fennell, R.D., Neely, R.B. (1983) 'The
    HEARSAY speech understanding system: an example of the
    recognition process', Proc. 3rd Int. Joint Conf. on AI,
    Stanford, Calif., pp.185-193.

Reed, T., (1988) 'Digital techniques for enhancement and
    classification of SEAMARC II sidescan sonar imagery`, JGR.

Reut, C., Pace, N.G., Heaton, M.J.P. (1985) 'Statistical
    properties of sidescan signals and the computer classification
    of sea beds´, Proc. Inst. of Acoustics 7, Pt. 3, pp.102-107.

Russell, G.T., Lane, D.M. (1986) 'A knowledge based system
    framework for environmental perception in a subsea robotic
    context`, IEEE Journal of Oceanic Engineering, OE-11, No. 3,
    pp.401-412.

Schlock, S.G., Leblanc, E. and Mayer, L. (1986) 'Sediment
    classification using a wideband frequency-modulated sonar
    system`, OTC Conference Proc., Houston, pp.389-398.

RECENT ADVANCES IN ACCURATE UNDERWATER MAPPING AND INSPECTION
TECHNIQUES

JOHN O. KLEPSVIK AND HANS OLAV TORSEN
Seatex A/S
Trondheim
Norway

ABSTRACT. The offshore oil and gas industry in the North Sea has set
new standards and led to increased demands for accurate and cost
effective subsea mapping and inspection techniques. The present paper
reviews some of the recent advances within this field and focuses on
different three dimensional (3D) mapping and imaging techniques. A new
class of high resolution wide-swath bathymetric systems is described
and compared. These systems requires positioning input of the highest
accuracy and consistency and the role of the Global Position System
(GPS) to meet these demands is discussed.

The paper puts special emphasis on the potential use of laser
scanning and laser radar techniques for high resolution 3D-imaging and
mapping. The basic criterias for use of optics subsea are discussed
and exemplified through the description of SPOTRANGE and SPOTSCAN
instruments as well as a new concept for a laser radar now being
developed at Seatex.

1.   INTRODUCTION

1.2 General Background

The spur to the recent advances within underwater mapping and
inspection techniques in the North Sea area stems mainly from the
development of offshore oil and gas resources in the North Sea.
Starting back in the late sixties, there has been more than 20 years of
continuous operations, oil field development, pipeline installations,
introduction of subsea production units, remotely operated techniques,
etc. in an area with rough environmental conditions, waterdepths
ranging typically from 70-400 m and partly very rough bottom topography
due to iceberg scourings and ploughmarks.

The size and importance of offshore oil and gas development is
probably best illustrated by the fact that the yearly average
investments (in Norway and UK) over the five last years are
approximately ten billion US$. And although seafloor surveying and
subsea inspection are not in the centre of the oil mans interests and
priorities, they still form an important part of the total picture. So

171

D. A. Ardus and M. A. Champ (eds.), Ocean Resources, Vol. II, 171–183.
© 1990 Kluwer Academic Publishers. Printed in the Netherlands.

offshore surveying which traditionally used to be a domain for governmental agencies and research organisations has become a rapidly developing industry in the North Sea, presently representing a 2-300 million $ market. The commercial demands from this market together with direct infusion of significant R&D money from the oil companies to the instrument developers (approximately 4-6 million $ yearly in Norway) have resulted in a variety of new sensors, systems and methods which are currently in its early application phase. Today is it probably a fair judgement to say that the commercial surveying industry in the North Sea Region is quite a bit ahead of the governmental and military survey agencies both in terms of methods, equipment, technical and operational expertise.

2. SEAFLOOR MAPPING

2.1 Some Typical Survey Requirements and Specifications

The ultimate results from a seafloor survey is normally a 3-D description of the bottom, presented as a contoured map. The quality of the map is again a function of the accuracy of the depth (z) measurement (sounding) and the position (x-y) of the depth measurement. A "rule of thumb" often applied is that depth accuracy should be 1/10 of position accuracy.

In the North Sea survey industry, a typical set of requirements and specifications for a detailed pipeline survey would be:

Survey depth	: Ranging from 50-300 m
Depth resolution	: 10 cm to 50 cm
Lateral resolution (footprint)	: 1 x 1 m up to 5 x 5 m
Survey corridor	: 1 x 200 km = 200 km^2
Positional accuracy	: 2-5 m
Survey speed	: Better than 4-5 knots
Area overlap	: 50-100%
Operational sea state	: At least sea state 5

These technical and operational requirements together with the demand for improved cost-efficiency have led to the development of several new high resolution wide-swath sonar systems as well as improved surface and underwater positioning systems, digital map processing techniques, improved tidal correction models, ray tracing models, etc.

CHAPTER 17

2.2 Examples of Wide-Swath Bathymetric Sonar Systems

In the following three different high resolution wide-swath systems are
described; systems which all, to some extent, can fulfil the above-
mentioned specifications but which technically or operationally are
quite different. Main system specifications are given in Table 1.

EM100. The EM100 is a shipborne, multibeam sonar system produced by
SIMRAD A/S, Norway. It has been on the market since 1985 and is today
the most utilized system for accurate seafloor mapping in the North
Sea.

Benigraph. The Benigraph is a prototype development made by Bentech
Subsea A/S, Norway. It is a very sophisticated, towed multibeam sonar
system which includes both a hydroacoustic and inertial positioning
system. The prototype version is fully operational and is being used
on commercial jobs with excellent results.

Bathyscan.
The Bathyscan manufactured by Bathymetrics, UK, is a so-called
topographic sidescan sonar or interferometric sonar which measures
range and angle (phase) to the reflecting object thereby generating
topographic 2-D information as the acoustical pulse travels along the
seafloor. Bathyscan is currently used for detailed seafloor mapping in
the North Sea, especially in shallower water.

TABLE 1. Wide-swath bathymetric sonar system.

Manufacturer	Type	Model	Frequency (kHz)	Range (m)	Max swath (m)	Beamconf. (no/x°x y°)	Range reso-lution (cm)	Max Prf (Hz)
SIMRAD	Multibeam	EM100	95	650	800	32/3°x2.5°	7.5	3
BENTECH	"	Benigraph	515	60	80	200/2°x1°	3	10
	"	"	740	50	70	200/1.5°x0.75°	3	10
	"	"	1000	40	60	200/1°x0.5°	3	10
BATHYMETRICS	Interf. sonar	Bathyscan 300	300	150	200	2/1°x50°	10	5

A comparison of the different systems involves a number of parameters,
however, in Seatex we have recently made a comprehensive study for a
major oil company where cost-efficiency has been calculated for a large
range of different mapping systems including those mentioned above.
Cost-efficiency is defined as how many km² processed map you get for
1.0 million US$. This study was based on the pipeline case mentioned
above. Some of the main conclusions are:

-   Shipborne multibeam systems are the most cost-effective,
    approximately 3 times better than the singlebeam echosounder and
    approximately 2-3 times better than the towed topographic sonar.

- The towed topographic sonar is approximately two times more cost-
  effective than the towed multibeam sonar.

- The most cost-effective system can produce up to 400 km² of
  digitally processed map per million US$.

3. POSITIONING

The positioning accuracy necessary to make a digitally processed map
with grid size of e.g. 5 x 5 m, must at least be better than 5 m. In
order to achieve such accuracies, the oil industry operates its own
surface radio navigation systems in the North Sea, typically two
classes:

- 2 MHz systems (over the horizon system), with range capabilities
  up to 3-400 km and accuracies ranging from 6-30 m. Such systems
  are Hyperfix, Argo, Spot, Geoloc.

- Line of sight systems, with range capabilities up to 100 km
  (depending on antennae height) and accuracies 2-10 m. Such
  systems are Syledis, Miniranger, Microfix, etc.

The TRANSIT satellite system is only used for positioning of fixed
platforms, or as one of many inputs in large integrated positioning
systems.
    The Global Positioning System (GPS) has been operational as a test
system for 6 years, but due to a limited number of satellites it gives
only approximately 2 x 3 hours of coverage per day in the North Sea
area. The system which is a military system operated by the US Air
Force, today gives an accuracy of approximately 30-50 m for civil users
(CA-code), with approximately one fix per second while the restricted
code (P-code) gives an accuracy of approximately 10 m and very high
dynamics. By the end of 1991 the GPS will have worldwide, continuous
3D coverage.
    In spite of poor coverage, the oil companies in the North Sea have
recognised the potential of the GPS as the future positioning system,
offering continuous, worldwide 3D coverage both for offshore and
onshore operation. The oil industry has therefore supported various
developments to improve accuracy, to adopt the system to the needs of
the offshore surveyor, to test and gain operational experience during
the GPS trial period. In Seatex we have been developing two
techniques, Differential Positioning Techniques (SEADIFF) and Phase
Measurement Techniques (SKYPHASE). The differential technique is shown
in figure 1 and is based on the assumption that the position errors
measured based on pseudo range on a fixed reference station, can be
used to correct the position measured on the mobile unit. The
corrections are then transmitted in real time to the mobile unit either
by separate broadcast or coded onto another radio navigation signal
(e.g. Argo, Hyperfix). The differential technique has shown that the
position accuracy can be improved by almost a factor of ten, from 50 m
rms to 5 m rms. The results of the differential technique is dependant

on the distance between the reference station and mobile unit
(baselength), however good results are achieved with baselengths up to
1000 km.

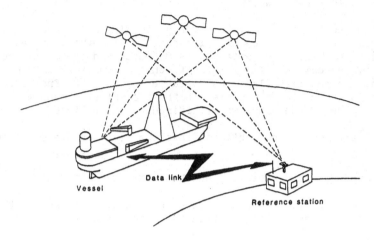

Figure 1. Differential GPS

The SkyPhase technique is currently based on the fact that mobile unit
is fixed for approximately 30 minutes while it takes measurements from
the same cluster of satellites as the reference. Typical achievable
accuracy is 1-2 PPM (parts per million) of baselength. With a
baselength of 10 km, this means 1-2 cm accuracy.
       The GPS opens up a revolution in positioning and navigation both
offshore and onshore. The big uncertainty just now is to what extent
the CA-code will be degraded for civil users. But even with more
degradation, there are ways to improve the system performance to the
extent that GPS seems to be the ultimate answer to our positioning
problems within EEZ.

4.   DETAILED INSPECTION TECHNIQUES

4.1 General Background

The offshore industry has also a great demand for detailed underwater
inspection and mapping techniques. Basically they need to visually
inspect their installations in order to verify its structural and
mechanical integrity, a requirement which normally has been satisfied
by use of underwater photography and TV, operated either by ROV's or
divers. Standard TV and photo are 2-D techniques giving little
quantitative information apart from the visual interpretation.
       As the offshore industry has moved into deeper water and seafloor
based production systems have come to replace traditional platform
solutions, there is a growing requirement for online 3-D imaging and
mapping techniques both for inspection purposes but even more as an
active tool during interventions, installations, docking operations,

levelling of structures, alignment of components, mating operations, accurate positioning, etc.

4.2 Optical Imaging in Water

A typical situation for active underwater imaging and surveying is shown in figure 2. The target of interest is illuminated by a light source, and the reflected radiation is detected and spatially processed (imaged) by the camera. In addition to the information-carrying radiation, the receiver also "sees" scattered radiation from dissolved particles and other inhomogenities in the transmission medium. This so-called path-radiance has two significant contributions.

1.   Backscattered source radiation into the receiver field-of-view (FOV).

2.   Forward scattered radiation due to multiple scattering.

The first contribution tends to reduce the overall contrast of the image, while the second contribution tends to reduce the spatial resolution.

Attenuation. Both the source radiation and the reflected radiation is exponentially attenuated due to the scattering process described above and to absorption in the transmission medium. A characteristic parameter is the beam attenuation coefficient c (m^{-1}). The inverse quantity, c^{-1} (m), the attenuation length, gives the distance to which the radiation intensity has fallen to $1/e$, or about $1/3$, of its original value.

The attenuation coefficient varies strongly with wavelength. Figure 3 shows the spectral characteristics of the attenuation coefficients. Attenuation of optical radiation is characterized by a narrow window in the blue-green part of the visible region. The window characteristic is especially pronounced for deep-ocean water, but less so for typical coastal and shallow water areas. Also shown is the beam attenuation for acoustic radiation.

The important fact that can be observed in figure 3 is that the attenuation coefficient for optical radiation in the blue-green part of the spectrum and acoustic radiation in the 1-5 MHz range have comparable magnitudes. The attenuation length has also recently been measured in the Northern North Sea during summer conditions. Figure 4 shows measured attenuation length variations visibility as function of depth for the green part of the spectrum. Below 100 m, the attenuation length varies from 10-16 m, indicating that considerable viewing distances are possible.

Principles of 3D-Imaging and Surveying. In order to obtain the (x,y,z) coordinates of an object relative to a reference-frame given by the orientation of the imaging instrument, we have to measure ranges and/or angles relative to our reference. The basic geometries involved are shown in figures 5 and 6. The range-azimuth geometry (a) has the advantage that only one observation position is needed, while for the

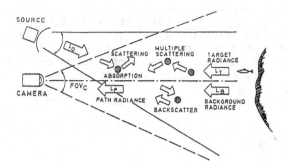

Figure 2.   Block diagram of basic imaging situation.

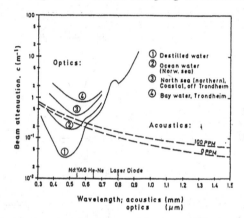

Figure 3.   Typical spectral attenuation coefficients for optical and
acoustical radiation.

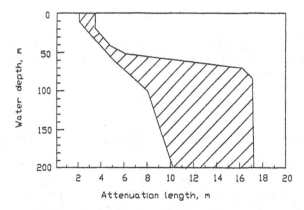

Figure 4.   Attenuation length as function of depth in Norwegian Coastal
Water.

range-range and angle-angle geometries we have to observe the object
from two positions in space. For the last two geometries, the
longitudinal resolution (Δz) strongly depends on the base-to-depth
ratio (b/z), which should be larger than approximately 0.1. The
computation of coordinates and estimates of accuracies only involves
elementary trigonometrical relations. In optics, angular information
is easily obtained using lenses. Range information can also be
obtained from time-of-flight measurement using pulsed lasers.

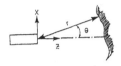

(a)   Range-azimuth

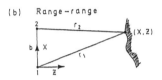

(b)   Range-range

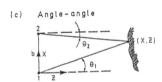

(c)   Angle-angle

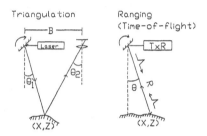

Depth accuracy vs range

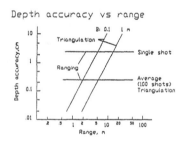

Figure 5.   Basic 3D imaging/   Figure 6.   3D imaging techniques.
            surveying geometries.   Depth accuracy vs range.
            (a) Range-azimuth
            (b) Range-range
            (c) Angle-angle.

4.3 Examples of Laser Based Subsea Inspection Instrumentation

SPOTRANGE. The SPOTRANGE instruments developed by Seatex A/S, combines
acoustics and a laser. Range measurement is done using acoustics, and
the laserbeam is coaxially aligned with the narrow acoustical beam and
provides an accurate visual definition of the target.
     SPOTRANGE can be used in combination with most TV-cameras; colour,
vidicon, CCD and SIT. The intensity of the laser beam is automatically

adjusted and modulated to ensure optimum definition of measurement
targets throughout the visibility range.

Two versions of the subsea unit are available; a single-unit model
employing a 2 mW He-Ne laser and a close range version featuring a
separate sensor-head with small dimensions for use on manipulators,
etc. (Fig. 7). The latter employs a 2 mW red laser diode as a pointer.
The specifications for the subsea units are given in Table 2.

Figure 7.   Photograph of SPOTRANGE subsea units, SR01 and SR02.

TABLE 2.   SPOTRANGE.   Specifications for the subsea-units.

Depth rating: 1000 m

Ranger (acoustic)	SR01	SR02
Frequency (MHz)	1.0	2.0
Range, max/min (m)	30/0.1	10/0.1
Resolution (mm)	5	3
Beamwidth (deg)	1.5	2.0
Laser pointer		
Wavelength (nm)[1]	633 (red)	670 (red)
Visible range (m)		
(Vidicon, CCD)	4-8[2]	3-4
Beamwidth (deg)	0.1	0.1
Dimensions/power		
Diam. (mm)	100	50
Length (mm)	375	80
Power (VA)	24/1.5	24/1.0

[1] Green (532 nm) is also now available with considerable increase in
visible range.

[2] About twice normal TV-range.

SPOTRANGE Applications. Until now SPOTRANGE has been used for accurate navigation of ROVs, metric scaling of video pictures by a grid overlay technique and accurate differential range measurements between two abigued objects.

The SPOTRANGE concept also opens up the use of triangulation and trilateration methods in subsea surveying, as depicted in figure 8. Other applications which are presently undertaken by Seatex A/S include:

- Laser guidance in docking operations: Installation of subsea modules using ROVs.

- Subsea levelling of templates and other bases for subsea construction.

- Continuous profiling of structural details for damage assessment, etc.

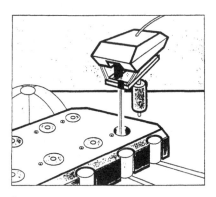

Figure 8.   Application of SPOTRANGE
in a docking operation.

SPOTSCAN.   Automated optical 3D systems, i.e. stereo-TV and laser scanners use the angle-angle geometry, and angular measurements are obtained using image-plane position sensors.   Stereo-TV (videogrammetry) is based on measurement of parallax-differences, and requires a fairly complex correlation processor as well as high contrast images in order to extract depth information.

Laser scanning techniques have been in use for several years in high-resolution printing applications.   Raster scanning of the laser beam is obtained by two scanning mirrors having a high resolution angular pick-off.   A more detailed description of the imaging geometry is shown in figure 9.   The laser spot is continuously observed by a camera using a position sensor as detector, and this combination provides the second angle, V_2, in the triangulation geometry.

The SPOTSCAN 3D imaging system, currently undergoing seatrials and industrialization at Seatex A/S, is based on the principles outlined above.   SPOTSCAN can be used both in a profiling mode and full 3D imaging mode and provides a real-time alternative to close-range

photogrammetry using stereo-photography. SPOTSCAN is also an imaging
instrument in the traditional sense, providing TV-pictures with a
comparable resolution, but with a limited framerate capability compared
to TV. However, in situations where the TV-system is backscatter
limited, SPOTSCAN can provide enhanced viewing ranges and image-
contrast due to the inherent backscatter discrimination properties of
the scanning method.

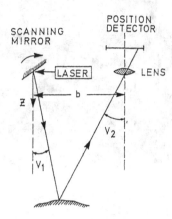

Figure 9.   SPOTSCAN 3D-laser scanner.
             (a) Measurement geometry.   (b) Pipeline profiling.

SPOTSCAN specifications are given in Table 3 and compared to stereo TV
(videogrammetry). The SPOTSCAN uses a new solid state green laser,
wavelength 532 nm, and output power of 20 mW which in relatively clear
seawater should give ranges of 5-8 m.

TABLE 3.   3D imaging systems.   Typical performance characteristics.

System	Spatial lateral (deg)	Resolution depth (mm)	Frame resolution (typical)	3D-Frame acquisition time (s)
Stereo-TV (videogrammetry)	0.1	1-3 (at 0.5 m)	256 x 256	10 - 20
Laser-scanner (SPOTSCAN)	0.1	0.1 - 0.5 (at 1 m)	240 x 180	3
Laser radar (RANGE FINDER)	0.1	30 - 50	64 x 64	0.5 - 1

The application areas for subsea inspection includes detailed
(sub-mm) corrosion-monitoring, crack-inspection, marine growth
assessment, fast pipeline profiling, etc., as well as shape and
dimensional control in general, as exemplified on figure 10.

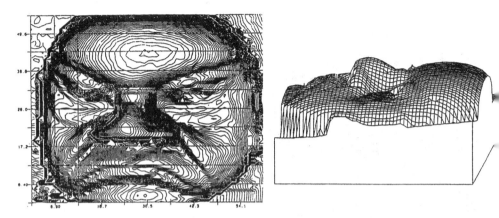

Figure 10.   3D contour map (left) and isometric projection generated by
SPOTSCAN.   Contour interval is 0.5 mm.

4.4 Future Developments

The SPOTSCAN continuous scanning principle is best suited for accurate
3D close up (1-3 m) inspection with submillimetric resolution, which
results in a compact and rugged instrument. For longer ranges, the
range-azimuth principle is favourable, using high energy, pulsed green
laser where range is derived from time of flight measurements. At
Seatex we have recently undertaken a comprehensive design study on a
subsea laser rangefinder utilizing Q-switched solid state green lasers
and fast optical detectors.
     The principle is shown in figure 6 and system configuration in
figure 11.
     Range capabilities for the optical range finder are determined by
laser pulse energy and the attenuation length in water. Based on
attenuation length figures given in figure 4 (Norwegian water), we have
predicted range capabilities for our new range finder and compared them
to existing instrumentations as shown in figure 12. The interesting
conclusion is that in relatively clear deep sea water, imaging ranges
up to 50-60 m will be possible, with depth resolution in the cm-range.

3D - Imager

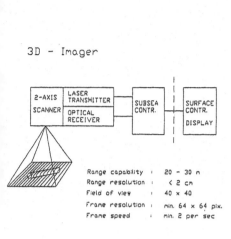

Range capability :   20 - 30 m
Range resolution :   < 2 cm
Field of view   :   40 × 40
Frame resolution :   min. 64 × 64 pix.
Frame speed   :   min. 2 per sec

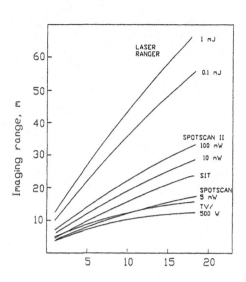

Figure 11. System configuration
for laser radar (range finder).

Figure 12. Range capabilities as
function of attenuation length.

5. CONCLUSION

The demand and requirements set forth by the offshore oil industry
operating in the North Sea, has resulted in a strong and continuous
development within accurate underwater and mapping techniques.

For seafloor mapping, high resolution wide-swath, multibeam sonars
have led to significant improvements both in terms of dataquality,
accuracy as well as cost-efficiency.

Positioning is an equally important parameter in the seafloor
mapping process as the depth measurement. The introduction of GPS as a
fully operational system from 1991 opens up a range of new applications
and techniques which will revolutionize marine positioning and
navigation.

For detailed 3D subsea imaging, the new laser techniques open up a
new dimension for accurate, real-time inspection and surveying methods
that can fulfill our dreams of mastering and exploring the underwater
world.

SUBSEA COMMUNICATIONS FOR SEMI-AUTONOMOUS ROVs

R.M. DUNBAR and D.R. CARMICHAEL
Heriot-Watt University
Edinburgh
Scotland
U.K.

Abstract. Through-water electromagnetic wave, optical, acoustic and
dispensed line signalling techniques are discussed, after a brief
review of baseband signals and noise, relevant to semi-autonomous
applications. A proposal for AROV intervention with the deep ocean
DUMAND array is outlined. The paper also contains a detailed account
of a technique of adaptive echo cancellation for the relevant problem
of high data rate acoustic communications in the presence of multipath
interference.

1.   INTRODUCTION

An autonomous submersible (AROV) is a complex system which depends on
artificial intelligence to enable it to carry out its mission without
intervention. However, in the real world there is often a desire to
communicate with the tetherless vehicle at levels ranging from status
interrogation to multi-function control and video transmission over a
through-water link. There are also requirements for cable-less
transmission of data between such vehicles and seabed sensors and
structures.
       The paper examines electromagnetic, optical, acoustic and
dispensed line techniques of signal transmission, particularly with
respect to bandwidth and range, with comment on multipath propagation
and noise limitations.
       The paper cites as a test case and as an example of technology
transfer the communication requirements for a tetherless submersible
designed to service the proposed DUMAND optical array at an ocean depth
of 4700 m.

2.   COMMUNICATION PARAMETERS

2.1. Baseband Signals

Sensor signals generally have low bandwidth requirements, of the order
of 10's to 100's of Hz. Video signals in real time on the other hand

185

D. A. Ardus and M. A. Champ (eds.), Ocean Resources, Vol. II, 185–199.
© 1990 Kluwer Academic Publishers. Printed in the Netherlands.

require several MHz and consequently are usually transmitted through
water by bandwidth-compressed slow-scan techniques, requiring seconds
to minutes for a frame update, (Fig. 1). Only an optical system
(Section 3.2) is capable of handling real time video.

Command and control signals are generally of low bandwidth
although the signals may be transmitted at the higher data rate
required by coding techniques employed to maximise reliability of
transmission.

Bandwidth requirements may be summarised as follows:

time multiplexed sensors	10's - 100's Hz
command and control	100's - 1000's Hz
slow-scan video	K's - 10 K's Hz
sonar	K's - 10 K's Hz
real-time video	2-10 MHz per channel

Format	Information Content (Bits/Frame)		Update Period 100k bits/s (s)	1k bits/s (hrs mins)	Bit Reduction Relative to Format 4
1. Conventional Colour TV	1.8	10^6	300	8 20	
2. Conventional Monochrome TV	600	10^3	100	2 50	
3. 512 x 512 pxls 8 bits/pxl	2	10^6	20	30	
4. 256 x 256 6 bits/pxl	400	10^3	4	7	
5. Frame Replenishment	$<40 \times 10^3 \rightarrow 400 \times 10^3$		.5→4	2→7	1:1→10:1
6. FR + DPCM	$<20 \times 10^3 \rightarrow 200 \times 10^3$		.25→2	1→2	2:1→20:1
7. 128 x 128 Reduced Image	$<5 \times 10^3 \rightarrow 50 \times 10^3$		.06→5	.25→5	8:1→80:1

Figure 1. Bandwidth reduction factors.

2.2. Operational Parameters

Signals degrade with range due to spherical spreading, absorbtion and
scattering. Water clarity imposes a limit an optical signalling
techniques. The presence of reflecting surfaces (water surface,
seabed, structural members, thermal layers) can produce acoustic
multipath interference, and prevailing noise provides an ultimate limit
on performance for all types of carrier, in terms of signal-to-noise
ratio.

2.3. Noise Levels

2.3.1. Acoustic Noise. Ambient acoustic noise spectra in seawater are
expressed in Figure 2 (Settery, 1987). The noise level decreases with
distance from the source and the attenuation rate varies with
frequency, as indicated by Figure 3. Consequently acoustic noise
levels in oceanic depths are very small except for very low frequency
natural and biological noise, short range higher frequency biological
noise, and man-made noise including sonar.

In some instances, particularly near to offshore structures in
continental shelf depths man-made noise can reach prohibitive levels.
Such is the case during piling operations on oil and gas platforms when
acoustic noise is generated over a wide frequency range at equivalent
source levels of kW/m² (Dunbar, 1987).

Multipath interference also provides an ultimate limit on
performance, in terms of signalling rate, and is a particular problem
where water depth is a small fraction of range, and in proximity to
reflecting surfaces.

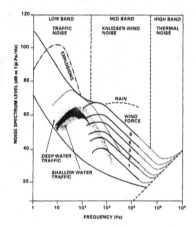

Figure 2.   Ambient noise
spectra in sea-water.

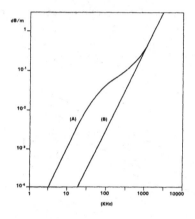

Figure 3.   Acoustic attenuation
coefficients
(A) sea water at 5°C
(B) fresh water at 5°C

2.3.2. Electromagnetic Noise.   Predicted levels of electric noise
fields are quoted in CCIR Report 322, for above-water applications
(CCIR Report 322, 1964), and these levels fall off rapidly with depth
due to refraction loss and exponential attenuation (Maxwell and Stone,
1963).   In deep water it is only locally generated noise sources that
are likely to produce significant signal levels at frequencies above
the kHz range.

2.3.3. Ambient Optical Noise.   Since underwater point-to-point optical
communication links are rather uncommon there is little in the
literature to quantify this problem.   Narrow band optical interference
filters are available for use with monochromatic light sources such as
lasers, enabling source-detector links to perform even in the presence
of daylight background 'noise'.

3. SIGNAL CARRIERS

3.1. Electromagnetic Wave Carrier

From the point of view of autonomous ROVs E-M wave signalling can
satisfy two requirements, (1) long range radio navigation for a
vehicle, with its antenna at or near the surface; and (2) short range
communications with another vehicle or with a seabed structure.

The attenuation of E-M waves in seawater is very severe
(Fig. 4) ruling out long range high frequency through-water signalling.
However kHz bandwidth signals can be transmitted over 10's of m from
magnetic loop (Fig. 5) or surface contour electric field antennas, this
being useful for communication with seabed structures in high acoustic-
noise environments.

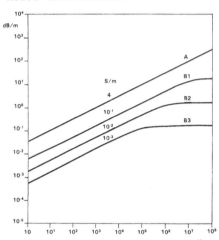

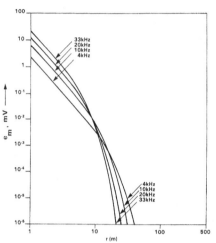

Figure 4. Electromagnetic wave
attenuation in (A) seawater and
(B1,2,3) fresh water.

Figure 5. Variation of e.m.f.
induced in loop antenna submerged
in seawater: computed values for
100 AT transmitter field.

3.2. Optical Carrier

The narrow "window" in the curve of E-M wave attenuation with respect
to frequency occurs around 10^{14} Hz and the minimum attenuation occurs
around 550 nm, the wavelength of green light (Fig. 6) (Myers, Holm and
McAllister, 1969). It can be observed that there is a shift towards
the blue end of the spectrum for the purest oceanic waters. This is
further illustrated in Figure 7, extracted from a DUMAND Proposal
document, 10th November 1982 (International DUMAND Collaboration,
1982). If one considers the attenuation length for oceanic waters as
25 m (Fig. 7) then communication should be possible over at least 250 m
for a 90 dB path loss. In addition, the optical carrier may be
modulated at 10's to 100's MHz, thus permitting real-time video
transmission through water (Dunbar, 1985). Such a system is actively
under investigation by the authors, for the DUMAND application.

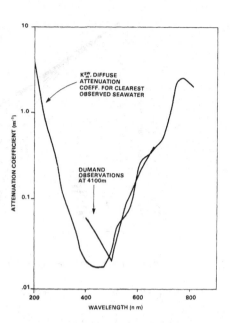

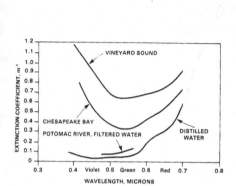

Figure 6. Spectral attenuation
coefficients in water (from
Myers, Holm and McAllister, 1969).

Figure 7. Attenuation versus
wavelength.

3.3. Acoustic Carrier

Acoustic carrier signalling is widely employed for through-water
communications and sonar and it is extensively documented, for example
Acoustical Oceanography, by Clay and Medwin (Clay and Medwin, 1977).
Reliable signalling is possible over surface to seabed links at rates
of 100's to 1000's bits/s, thus satisfying most command, control, data,
and slow-scan T.V. requirements for AROVs (Borot and Brisset, 1987).
If higher signalling rates are required, e.g. for faster video frame
update, then ranges become limited because of the necessary higher
carrier frequencies (Fig. 8).
     A serious limitation to high data rate acoustic signalling is
multipath interference. This is discussed in detail in Section 5,
together with an adaptive echo cancellation technique designed to
combat this type of interference. Special forms of modulation have
also been proposed to alleviate this problem (Ayela and Le Rest, 1987).

3.4. Dispensed Line Signalling

An AROV employing a dispensed line communication link is more strictly
a tethered vehicle. However, since the hydrodynamic drag aspect of a
normal umbilical cable is virtually absent with a fine dispensed line
the AROV is essentially as free to manoeuvre as in a truly tetherless
mode.

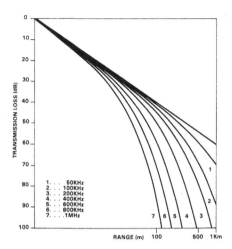

Figure 8.   Acoustic transmission loss in seawater.

Three main types of dispenser systems are applicable.

(1)   a single insulated, twin or quad transmission line, with a
      diameter of the order of 2 mm: for example, a bandwidth of 10-20
      kHz is possible over lengths of several km with a twin wire
      system.

(2)   a miniature coaxial cable: such a cable, of diameter around
      2.5 mm, is employed by a mine countermeasures vehicle to transmit
      command and video signals over at least 1 km.

(3)   an optical fibre: such a system holds the greatest promise as
      regards bandwidth, immunity to interference, and zero
      electromagnetic signature; ruggedised optical fibres with very low
      attenuation (< 3dB/km) are being produced for this application and
      re-reeling systems are under development to enable the fibre to be
      used on several missions.

All dispensed line systems suffer from the potential hazards of
snagging on an obstacle and breakage and AROVs employing such a
technique must be designed with tetherless location or homing
intelligence, at the very least.

4.   TEST CASE

There is a significant possibility that an AROV could be involved in
the deep ocean DUMAND project. Tethered ROVs have been proposed in an
intervention role from the early stages of the DUMAND project
(Gundevson, 1978) but an AROV has significant advantages from the point
of view of potential entanglement.

4.1. The DUMAND Project

DUMAND (Deep Underwater Muon and Neutrino Detector) is a project to build a deep underwater unmanned laboratory for the study of (1) high energy neutrino astrophysics, (2) cosmic ray physics, (3) neutrino physics above ITeV, and (4) geophysics and ocean science. The proposed site for the laboratory or measurement facility is approximately 25 km west of Keahole Point on the Island of Hawaii, at a depth of 4.7 km.

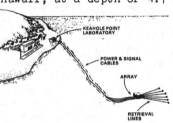

Figure 10. Disposition of the DUMAND detector at 4.7 km depth in subsidence basin approx. 25 km off Keahole Point, Island of Hawaii.

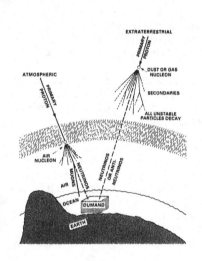

Figure 9. The concept of the DUMAND experiment.

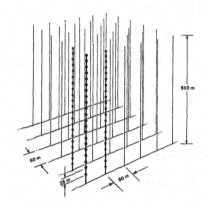

Figure 11. The 1982 concept of the DUMAND array. The 1988 concept is an octagonal array.

The overall concept is illustrated in Figure 9, the disposition of the DUMAND detector in Figure 10, and the original proposal (1982) for the DUMAND optical array in Figure 11. The original proposal was to instrument a volume of water 250 x 250 x 500 m^3 on the ocean floor, with strings of photo detectors, the outputs of the photo multiplier tubes being taken to shore via fibre optic cables for signal processing

and analysis. The current plans (Stenger, 1988) for the proposed
DUMAND array differ from the 1982 proposal but they still involve a
large number of photo detector modules on strings.

4.2. Intervention by AROV

A tetherless or string-reliant AROV could visually examine the entire
array, including photo multiplier tubes (PMT), transmitting the video
via a modulated laser in real-time to a nearby active PMT, or dedicated
photo detector. Illumination is required for the TV camera so a
compromise is needed between low light level camera requirements and
saturation of a de-sensitised PMT. It would be advantageous to use
illumination outside the spectral range of the PMTs, to minimise the
possibility of saturation. The peak sensitivity of the PMTs is around
400 mm and the response to red light around 650 mm would be very small.
Video transmission could then take place over a laser link using a
device in the green/blue part of the spectrum; for example a frequency-
doubled Nd:YAG laser at 532 nm, or a frequency doubled semiconductor
diode array.
     Command and control signals could be transmitted over an optical
link from a fixed seabed installation, but it is more likely that a
high data rate acoustic link, transmitted from a fixed point within or
near to the array, would be a simpler and more cost effective
technique.
     A semi-autonomous ROV system could provide solutions to many of
the deployment, inspection and other intervention requirements of the
DUMAND array, and the existence of photo-detectors would provide a
presently unique opportunity for real-time video communications with an
AROV. The more complex problem of wideband communication between two
independently manoeuvring AROV`s via a tracked optical link is
currently under investigation (Dunbar and Tennant, 1987-1989).

5.   MULTIPATH ACOUSTIC INTERFERENCE AND ADAPTIVE ECHO CANCELLATION

This section outlines in detail a possible receiver structure suitable
for use on a short range multipath channel. The channel is
characterised for multipath and non-multipath conditions and an attempt
is made to estimate error rates and transmission ranges for a practical
communication system.

5.1. Channel Limitations

For a short range, high data rate underwater acoustic transmission
system the two main factors affecting error rate at the receiver are
multipath interference and the signal to noise ratio. Other factors
are present but their effects are minimal. For example, dispersion of
the transmitted waveform over a 200 m path will typically be of the
order of 5 μs and is therefore negligible in comparison with the bit
period if data rates in the region 10 k bits/s are used. Doppler shift

of the carrier is 0.03% f_c (where f_c = carrier frequency) for a relative transducer motion of 1 m/s and so is also negligible.

Factors affecting the signal to noise ratio on an echo free channel are:-

1. Transmitted acoustic power   (W)
2. Transducer beamwidth   (Z)
3. Transmission loss   (dB)
4. Noise power   (W)

The maximum acoustic power delivered to the water is limited by the cavitation threshold and the transducer element area.

Transmission loss is governed by spherical spreading and absorption due to the bulk viscosity of the water and is given by

$$TL = 20 \log \left(\frac{R}{R_o}\right) + \alpha(R-R_o) \text{ dB re } 1\mu Pa$$

$$\text{where} \quad R = \text{transmission range (m)}$$

$$R_o = 1 \text{ m}$$

$$\alpha = \text{absorption coefficient}$$

$$= .19 \text{ dB/m at } f_c = 600 \text{ kHz}$$

The source level is the pressure level at 1 m from the transmitter and is defined as

$$SL = 170.8 + 10 \log Pa + DI_T \text{ dB re } 1 \mu Pa$$

$$\text{where Pa} = \text{total acoustic power (W)}$$

$$DI_T = \text{directivity index of the transmit transducer (dB)}$$

5.2. Noise

Sea noise characteristics given (Horton, 1957) indicate that the noise is predominantly thermal for frequencies above 200 kHz. The noise power spectral density may be approximated by

$$NL_o = -75 + 20 \log f_c \text{ dB re } 1 \mu Pa/Hz$$

$$\text{where } f_c = \text{frequency (Hz)}$$

The noise is isotropic and may be discriminated against by using directional transducers. If the noise is assumed to be Gaussian and the signal bandwidth is small in comparison with f_c then the noise at the receiver may be approximated by

$$NL = -75 + 20 \log f_c + 10 \log B_w - DI_R \text{ dB re } 1\mu Pa/Hz$$

$$\text{where } B_w = \text{signal bandwidth (Hz)}$$

$$DI_R = \text{directivity index, receiver}$$

The signal to noise ratio at the receiver may be written as:

$$S/N = SL - TL - NL \text{ dB}$$

5.3. Power Limitations

The maximum acoustic power which may be delivered to the water by a transducer is limited by conversion efficiency and cavitation. The cavitation threshold is frequency dependent and has been given (Urick, 1975). It has been measured at between 150×10^5 and 380×10^5 Pa for shallow water at $f_c = 550$ kHz. In practice transducer operation will be limited by electrical losses well before the onset of cavitation.

Fig. 12 shows S/N against range for different values of transmitter power and different transducer beamwidths.

As may be seen the use of narrower beamwidths significantly increases the S/N whereas increase in output power has less effect.

If an S/N of 10 dB at the receiver is taken as the limit for acceptable communication then a range of 350 m may be obtained using transducers with BW = 34° and a transmitted power of 10 W. If transducers with BW = 6° are used then the range may be extended to 500 m for the same power.

In practice, where a moving vehicle is used, the transmit and receive transducers must be aligned and so a compromise must be sought between ease of tracking and beamwidth. A typical system might have a range of approximately 400 m.

5.4. Error Rate

If coherent PSK is employed as the modulation type then matched filter detection will give a received error probability of $Pe = 10^{-6}$ for S/N = 10 dB as indicated in Figure 13.

If Nyquist pulses $(\frac{\sin x}{x})$ are used as the baseband pulse shape then a data rate of 10 k bits/s would require a double sided transducer bandwidth of 10 kHz. Transducers with a bandwidth of 200 kHz could therefore support a transmission rate of 200 k bits/s. In practice, however, the required baseband pulse shape and the matched filter are difficult to implement and it is more common to use a rectangular pulse shape for which the matched filter becomes an integrate and dump unit. The bandwidth in this case is greater than that required for Nyquist pulses.

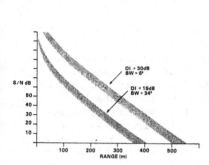

Figure 12.   S/N ratio v.
transmission range.

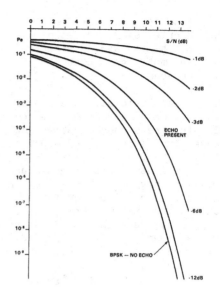

Figure 13.   Error probability v.
direct signal to noise ratios for
different echo magnitudes.

Performances curves for the spectral occupancy of binary PSK
signals have been given (Prabhu, 1976) for different types of baseband
pulse.   It is found that for a data rate of 30 k bits/s the use of
rectangular pulses will result in only ≈2% of the signal power being
contained outside a double sided passband of ± 200 kHz.
Thus for a practical communication system data rates of up to
30 k bits/s may be supported with currently available transducers
giving a range of approximately 400 m for a S/N of 10 dB at the
receiver.

5.5. Multipath

Multipath interference caused by specular reflections from boundaries
such as the sea surface and bottom also contribute significantly to the
error rate on the short range acoustic channel.   Figure 13 shows the
symbol error rate for a PSK signal corrupted by Gaussian noise and a
single point source echo.   The worst case error probability is plotted
against direct signal to noise ratio for different echo amplitudes.
As can be seen the presence of a single echo may considerably
degrade the error performance of the system.   For example a channel
with S/N = 10 dB at the receiver will suffer an error rate
deterioration from $Pe = 10^{-6}$ to 10^{-2} in the presence of a single echo
of strength -3 dB relative to the direct signal.
Thus the prime consideration for a suitable receiver structure
must be the reduction of multipath interference in the received signal.
A practical receiver for use in a multipath environment must have the
following functions.

1. Echo removal
2. Carrier synchronisation
3. Bit timing synchronisation
4. Automatic gain control
5. Decoder

5.5.1. Echo Removal. A decision fed back equaliser (DFE) is used to identify echo components and remove them from the input signal, thereby allowing the synchronisation and gain control algorithms to perform correctly. This type of equaliser structure is used since it is known to be particularly effective for channels with severe distortion.

The general equaliser structure is shown in Figure 14. The equalised signal is the sum of the feed forward and feed back parts of the equaliser. The feed forward part acts as a matched filter for the pulse shape which in this case is rectangular. Bandwidth limiting imposed by the transducers and the channel is negligible for data rates in the order of 10 k bits/s and so the forward half of the equaliser may be replaced by an integrate and dump unit synchronised to the bit timing interval.

The feedback tap weighting coefficients are adjusted using the LMS stochastic gradient algorithm which results in minimum mean square error at the equaliser output.

The most significant problem associated with the DFE structure is due to error propagation caused by incorrect decision values propagating down the feedback delay line. This will cause misadjustment of the tap weighting values. The problem is not catastrophic for isolated bit errors but may prevent equaliser convergence if the other control loops have not synchronised sufficiently. It is therefore beneficial to ensure as much independence between the control algorithms as possible.

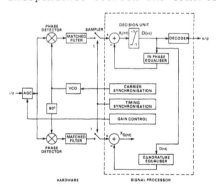

Figure 14. Decision feedback equaliser.

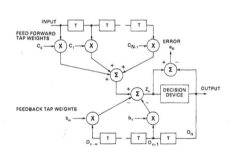

Figure 15. Adaptive receiver unit.

5.5.2. Carrier Synchronisation. The carrier recovery loop is implemented as a discrete version of the conventional Costa's loop with a hard decision in the in-phase channel.

The error output from the phase detector is given by

$$e_{c(nT)} = D_{(nT)} \cdot X_{q(nT)}$$

where $e_{c(nT)}$ = carrier loop error at $t = nT$

$D_{(nT)}$ = decision level

$X_{q(nT)}$ = quadrature phase i/p value

5.5.3. Timing Synchronisation. The error algorithm used is that proposed by (Gardner, 1986) and is given by

$$e_{T(nT)} = X_{I(n-\frac{1}{2})T}[X_{I(nT)} - X_{I(n-1)T}] + X_{q(n-\frac{1}{2})T}[X_{q(nT)} - X_{q(n-1)T}]$$

where $X_{I(nT)}$ = in-phase i/p value

As may be seen the algorithm is not decision directed and therefore offers a degree of independence from decision errors associated with the equaliser structure. The algorithm is also phase independent and so the timing aquisition loop is not affected by the state of the carrier recovery loop. The main disadvantage with this algorithm is that in-phase and mid-phase timing values are required which necessitates the use of two taps/bit in the equaliser. The receiver must sample at twice the bit rate and the required equaliser tap length is effectively doubled. It is felt, however, that the improvement in performance of the complete receiver afforded by the use of this algorithm is worth the increase in complexity.

5.5.4. AGC. The gain error is given by

$$e_{G(nT)} = X_{I(nT)}^{2} + X_{Q(nT)}^{2} - 1$$

The gain control loop is also phase independent. The AGC unit is implemented as a multiplying DAC and is driven from the output port of the signal processor.

5.5.5. Decoder. It is necessary to employ differential encoding of the data stream in order to overcome the 180° phase ambiguity inherent in the carrier recovery loop. Some form of error correction coding might also be usefully employed within the system giving improved error performance at the expense of effective signalling rate.

5.6. System Summary

A schematic diagram for the prototype receiver unit is given in Figure 15.

The system consists of an analogue input section followed by a signal processor. The input is first passed through the AGC which is software driven from the processor. All control algorithms are executed on a TMS 32010 signal processor. The output from the AGC is fed to a quadrature oscillator to produce in-phase and quadrature baseband components. These are then passed through matched filter detection units and sampled to produce input values for the signal processor. The quadrature equalisation is performed by the processor and the system output is derived from the decoder section.

Software simulation of the complete system has been carried out and results indicate that satisfactory convergence of the control loops is obtained from an arbitrary unlocked condition. Convergence may take place in the order of a few hundreds of bits depending on channel conditions such as signal to noise ratio and echo strength. Simulation also suggests that lock is generally maintained under most channel conditions though carrier fading can be a severe problem particularly in a changing echo environment.

The prototype unit is currently under development, nearing completion, and it is expected that a practical assessment of system performance will be achieved in the near future.

6.   ACKNOWLEDGEMENTS

The authors wish to acknowledge the contribution made to this area of research over several years by colleagues, in particular I.T. Anderson, A.M. Dobson, S.J. Roberts, A. Settery and A.W. Tennant. The authors also wish to thank the Hawaii DUMAND Centre for their willingness and openness in discussing their ideas regarding a deep ocean optical array.

The work described in this paper has received support from the U.K. Science and Engineering Research Council and the Ministry of Defence.

7.   REFERENCES

Ayela, G., Le Rest, S. (1987) "A new self-contained mutltipath protected acoustic transmission system for offshore environment", Proc. Deep Offshore Technology, 4th International Conference, paper V.7b, pp.86-97.

Borot, P., Brisset, L. (1987) "ELIT - a new underwater unmanned untethered vehicle", Proc. Deep Offshore Technology, 4th International Conference, paper V.7a, pp.73-85.

CCIR Report 322 (1964) "World distribution and characteristics of atmospheric radio noise", I.T.U., Geneva.

Clay, C.S., Medwin, H. (1977) "Acoustical Oceanography", John Wiley & Sons, Inc.

Dunbar, R.M. (1985) "Remote inspection, maintenance, and repair of the DUMAND array", Hawaii DUMAND Center Report HDC-2-85, University of Hawaii.

Dunbar, R.M. (1987) "Electric, magnetic and acoustic noise generated underwater during offshore piling operations", Proc. Inst. Acoustics, Vol. 9, Part 4, pp.12-17.

Dunbar, R.M., Tennant, A.W. (1987-1989) "Optical communication link for deep ocean autonomous remotely operated vehicle", SERC Ref. GR/D/8657.7, 31.03.87 to 30.09.89.

Gardner, F. (1986) "A BPSK/QPSK timing error detector for sampled receivers", IEEE Trans. Comms., Vol. COM-34, No. 5, pp.423-429.

Gunderson, C.R. (1978) "Maintenance of the DUMAND array", DUMAND Ocean Engineering Workshop, Hawaii DUMAND Center, University of Hawaii, pp.161-169.

Horton, J.W. (1957) "Fundamentals of Sonar", United States Naval Institute, Annapolis.

International DUMAND Collaboration (1982) "Proposal to construct a deep-ocean laboratory for the study of high-energy neutrino astrophysics, cosmic rays and neutrino interactions", Hawaii DUMAND Center, University of Hawaii.

Maxwell, E.L., Stone, D.L. (1963) "Natural noise fields from 1 c.p.s. to 100 kc", IEEE Trans. Ant. and Prop., Vol. AP-11, No. 3, pp.339-343.

Myers, J.J., Holm, C.H., McAllister, R.F. (1969) "Handbook of Ocean and Underwater Engineering", McGraw-Hill Book Co., New York, pp.3-24.

Prabhu, U.K. (1976) "Spectral occupancy of digital angle modulation signals", B.S.T.J., 55, No. 4, pp.429-452.

Settery, A., op.cit., Figure 3.1.

Settery, A. (1987) "High data rate acoustic signalling for through-water video transmission", MSc thesis, Heriot-Watt University.

Stenger, V.J. (1988) "DUMAND: Progress and Status", Report, HDC-3-88, Hawaii DUMAND Center, University of Hawaii: presented at Neutrino '88, Tufts University.

Urick, R.J. (1975) "Principles of Underwater Sound", McGraw-Hill Inc., Second Edition.

TECHNOLOGY TRANSFERS IN UNDERWATER ACOUSTICS
COMMONALITY BETWEEN OCEAN AND NAVAL SYSTEMS APPLIED IN UNDERSEA WARFARE
AND HYDROCARBON MARKETS

PIERRE SABATHE*

and

HANS-JÖRGEN ALKER†

1. INTRODUCTION

Underwater acoustic systems have significantly evolved during the last decade, both in technology and application. Naval and offshore requirements are important factors in the definition of the next generation of systems and a significant technology transfer is expected into other sectors.

In the Naval sector two kinds of minehunting systems have been designed and developed, one for clearing mine fields, the second for harbour channel surveillance.

In both these systems, state of the art technology and technique to detect and locate mines have generated high performance systems: sonars with narrow beams, one tenth of a degree, and high range definition, around 0.2 m. Search for sunken objects and identification of navigation obstacles have been their first civilian applications.

Technologies developed for military active sonars and subsea navigation systems are widely applied by multibeam echo sounders for bathymetric mapping and by sidescan sonars for bottom imaging. Also, naval surveillance systems have potential for "on-site" surveillance of the oceans.

2. BASIC TECHNOLOGY IN MINEHUNTING

The two main active mine countermeasures are minesweeping and minehunting. In minehunting, mines must be found before disposal. The minehunting sonar is designed for this purpose. Bottom mines must lie at depths less than 100 m to be effective against their surface target. Finding a mine means the ability to recognise its size and shape among mine-like objects such as rocks or containers. The sonar must give an unambiguous image of the mine at a distance great enough for the minehunter safety, i.e. 150 m.

* Thomson Sintra ASM, Cagnes Sur Mer, France.
† Simrad A.S, Horten, Norway

D. A. Ardus and M. A. Champ (eds.), Ocean Resources, Vol. II, 201–214.
© 1990 Kluwer Academic Publishers. Printed in the Netherlands.

The bottom mine is a cylinder two metres long and half a metre in diameter which when reflecting ultrasound provides echo highlights. In order to separate them, even when arriving simultaneously, the minehunting sonar must receive them on different adjacent beams: about one tenth of a degree and very low side lobes for good contrast. In the same way, range resolution must be around 0.2 m. This is possible by transmitting either a very short pulse, or better a frequency modulated pulse combined with pulse compression in the receiver.

Another feature improves mine recognition, its shadow: sonar pulses are reverberated by the bottom, except behind the mine which is masked from the sonar. This lack of echo following the mine echoes is called the shadow and as the sonar is looking at a grazing angle the shadow is larger than the mine thus magnifying the mine shape.

The quality of the mine image also depends on the signal to noise ratio. Ping to ping correlation is used to improve it, applying special image processing to remove the minehunter random movement due to waves.

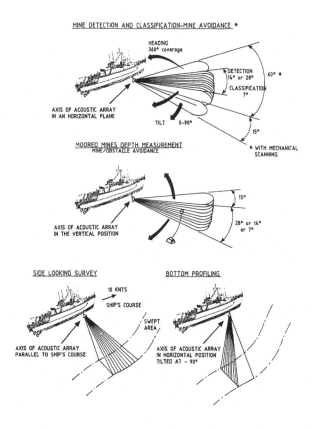

Figure 1.   TSM 2022 in operation

Minehunting sonar techniques, using narrow beams, can be applied to other projects simply to reorientating the transducer array as shown in Figure 1 for the THOMSON SINTRA TSM 2022 minehunting sonar.

- With the beams spread in a vertical plane it is possible to measure the target height over the bottom, for example the depth of a tethered mine.

- With the beams orientated to the bottom, it provides a multibeam echo sounder function.

- One narrow beam directed on one side of the ship gives a sidescan sonar. A specific type of minehunting sonar is used in minewarfare for surveillance, i.e. for a permanent reconnaissance of the channels providing access to harbours so as to guarantee the absence of mines within them.

3.   COMMONALITY BETWEEN OCEAN SYSTEMS AND MINEWARFARE SONARS

The commonality between ocean systems and minewarfare sonars are best illustrated by the following three examples:

- sidescan sonar

- multibeam echosounders

- multibeam for underwater vehicles

3.1. Sidescan Sonars

Two typical systems developed by THOMSON SINTRA ASM are:

- the French Navy minehunting sidescan sonar DUBM 41 for shallow water operations.

- the IFREMER deep water advanced acoustic imaging towed system SAR.

Their main characteristics are given in Table 1. The purpose of the systems are different:
The DUBM 41 must enable classification of mines on a single run. In order to achieve a constant resolution, the height over the bottom is fixed: 6,5 m and the beam shape is specifically designed.
As seen in Figure 2 the sonar images are excellent. The sunken aircraft (a WW II B 17) appears very clearly.
The SAR is designed for bottom survey at great depth. The speed is limited by the towing cable (length: 8,500 m) and a sufficient coverage is provided by the increased sonar range. The height over the bottom is adjustable and the fish is also fitted with a subbottom profiler operating at 3.5 kHz.

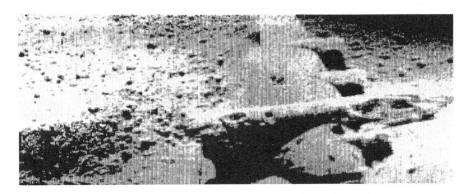

Figure 2.   A World War II bomber wreck displayed by Thomson DUBM 41.

Figure 3.   Abyssal slope at depth from 4800 m to 5300 m displayed by
Ifremer/Thomson SAR.

Images, as seen in Figures 3 and 4 are very interesting giving a
good mapping of the bottom and striking images of wrecks. For the
Titanic as for the mines the shadows magnify the outlines and allow a
quick recognition.
        The sonar raw data is fully recorded on each sonar:

-       magnetic recording on DUBM 41 for onshore treatment allow
        comparison of consecutive surveys for detecting the newly laid
        mines;

-       numerical optic disk recording on SAR for onshore process used
        with an interactive software, called TRIAS (Traitement des Images
        Acoustiques des Sonars latéraux), offers many processing
        possibilities:

* geometric corrections (slant range/speed/slope/yaw/bathymetry),

* contrast enhancement,

* image display at various scales (black and white/colour),

* mosaic mapping.

Figure 4.   Titanic's bow and its shadow displayed by Ifremer/Thomson
SAR.

As indicated in Table 1, DUBM 41 area coverage is smaller than
SAR's.   THOMSON SINTRA ASM is developing a new surveillance sonar for
the French Navy, called DUBM 42 for which the range towing speed and
operating depth (Table 2) is improved, thus the covered area is
multiplied by 10.   In order to maintain a full coverage with good
resolution the DUBM 42 is a multibeam sidescan sonar, which represents
the next step in the evolution of the shallow water sidescan sonars.

3.2. Multibeam Echo Sounders

Multibeam echo sounders are essential for effective seafloor mapping
(Fig. 6) in shallow and deep waters.   They are based on the crossed
beam technique: wide and flat single beams for the transmitting array,
narrow and multibeams for the receiving arrays as shown in Figure 5.
In Table 3 multibeam echo sounder characteristics are compared with
those of the minehunting sonar.
Shallow water echo sounders operate on continental shelves at
speeds twenty times higher than conventional equipment.

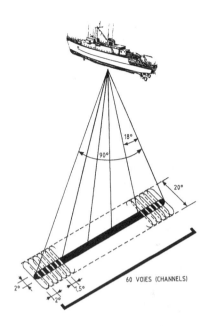

Figure 5.   Echo sounder beams.

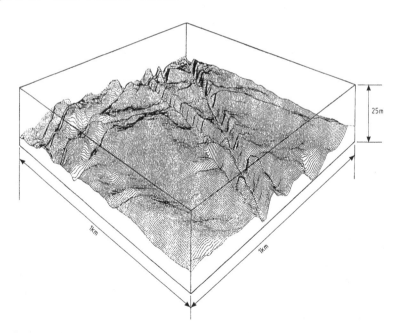

Figure 6.   Bathymetric seabed mapping with multibeam echo sounder (EM-100).

TABLE 1. DUBM 41 and SAR Characteristics

CHARACTERISTICS	DUBM 41	SAR
OPERATING DEPTH	100 m	6000 m
SURVEY SPEED	4 knots	2 knots
RANGE	2 x 50 m	2 x 750 m
AREA COVERED PER HOUR	.74 km²	5.56 km²
RANGE RESOLUTION	.07 m	.50 m
HORIZONTAL BEAM WIDTH	.12°	.5°
LATERAL RESOLUTION AT MAXIMUM RANGE	.1 m	6.5 m
HEIGHT OVER THE SEA BOTTOM	6.5 ± .5 m	ajustable around 70 m

TABLE 2. DUBM 41 and DUBM 42 Characteristics

CHARACTERISTICS	DUBM 41	DUBM 42
OPERATING DEPTH	100 m	300 m
SURVEY SPEED	4 knots	10 knots
RANGE	2 x 50 m	2 x 200 m
AREA COVERED PER HOUR	.74 km²	7.4 km²

Deep water echo sounders can be used for all deep sea measurement requirements and operate at speeds sixty times higher than conventional echo sounding equipment. These echo sounders have also the capability of logging the raw data for instant replay or off line processing.

3.3. Multibeam Sonars for Underwater Vehicles

The utilization of underwater vehicles depends on sensor capabilities and limitations, and has been boosted by advanced sensors developed for naval applications. Reduction of weight/space/power has opened up a market-niche for compact, on-board instrumentation. High-frequency sonars (100 kHz to several MHz) with small-sized transducers, sector-resolution capabilities and "image" -enhancement techniques are developed for close-area search/localization missions. The technology is now applied to seabed imaging/profiling and obstacle-avoidance on general-purpose vehicles in a variety of ocean applications.

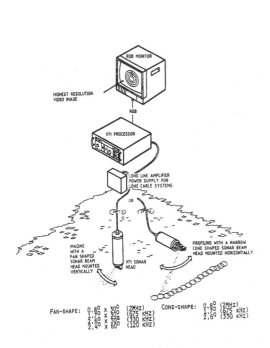 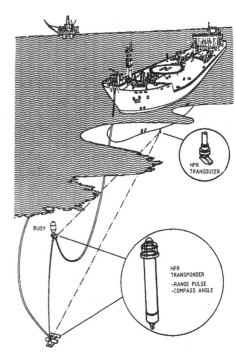

Figure 7. Vehicle configuration   Figure 8. Special-purpose
of high-resolution sonar.   subsea positioning and control.

The integrated solution for vehicle operation is given in Figure 7. Vehicle sonars are configurable with exchangeable sonarheads, selection of transducer beam-shapes and 1- or 2- axis transducer drives. Fast-scan implementations employ a combination of electronic beamforming and mechanical training. High-resolution capabilities in comparison with surface operation are given in Table 4.

TABLE 3. Multibeam echo sounders characteristics compared to minehunting sonar TSM 2022.

CHARACTERISTICS	MINEHUNTING SONAR TSM 2022	LENNERMOR TSM 5260	NADOZMOR TSM 5265
MAXIMUM DEPTH	> 500 m	> 500 m	> 10,000 m
TRANSVERSE COVERAGE	.5 water depth	2.4 x water depth	2 x water depth
ANGULAR RESOLUTION	.3°	5°	2°
DISTANCE RESOLUTION	.3 m	.3 m	1 m
NUMBER OF BEAMS	60	20	60
OPERATING FREQUENCY	200 kHz	100 kHz	12 kHz
MAXIMUM SPEED	10 knots	15 knots	15 knots

TABLE 4. Multibeam vehicle sonar characteristics compared with minehunting sonar TSM 2022

CHARACTERISTICS	MINEHUNTING SONAR TSM 2022	VEHICLE SONARS	
		MESOTECH 977**	TSM 5451
MODE OF OPERATION	SURFACE	SUBSEA 1000 m max.	SUBSEA 300 m max.
ANGULAR RESOLUTION	0.17°*	0.225°	0.5°
DISTANCE RESOLUTION	15 cm*	15 cm	10 cm
SECTOR COVERAGE	7°* MECH.TRAINABLE	7.4 km² MECH.TRAINABLE	48° ELEC.TRAINABLE
NUMBER OF BEAMS	60	32	96
OPERATING FREQUENCY	200 kHz	330 kHz	570 kHz

*   CLASSIFICATION MODE
**   MESOTECH 972 IS THE SIDESCAN VERSION

SIMRAD has developed a family of compact sonar systems applicable to vehicle operations. The technology has found its use in complementary applications like:

-   Ice detection, under ice mapping.

-   Compact close-area surveillance sonars with fixed seafloor transducer installations.

-   Trawl headrope sonars for fishery/fishery-research producing information of fish-gear dynamic behaviour.

The THOMSON TSM 5451 is devoted to underwater object relocation and mine neutralization.

4.   TECHNOLOGY APPLIED IN HYDROCARBON MARKETS

The offshore markets are highly active in the search for high performance/reliable instrumentation and have through R & D instruments made a significant "technology push" and created new and innovative acoustic systems. Within application areas like:

-   Subsea positioning and navigation,

-   Subsea monitoring and control,

instrumentation has gone through several development cycles (product generations) resulting in a qualified and mature technology.
New applications for acoustic instrumentation involve:

-   Deep water position-reference systems (> 1000 m, 1% slant-range resolution). Integrated position/control systems for offshore oil/gas loading.

-   Long-distance (>5 km) control of subsea templates, subsea sensors for gas leakage and pipeflow monitoring.

-   Streamer-tracking systems for airguns, multiple streamers and streamer separation.

-   Continental shelf, high resolution seabed mapping.

5.   COMMONALITY BETWEEN SYSTEMS

Commonality can be demonstrated by the following applications:

-   Underwater Position Reference systems in deep sea mining.

-   Acoustic data-transmission systems in subsea monitoring and control.

5.1. Underwater Position Reference Systems

During the mid-seventies, the first generation of subsea positioning
and navigation systems for oil-rigs and diver-support vessels were
developed. By applying the measurement technique of super-short
baseline (SSBL), an accurate position fix based on range/bearing-
estimation is established between a reference point (transmitting/
receiving transducer) and a single transponder.

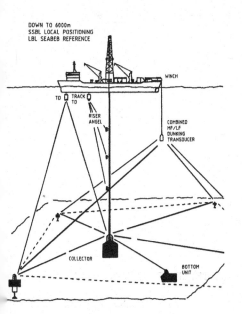

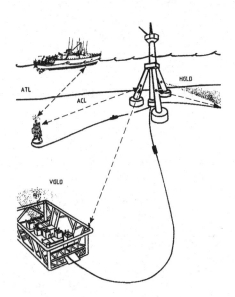

Figure 9.   Subsea positioning   Figure 10.   Integrated solutions
applied in deep-sea mining   for subsea monitoring and control

     Further development of surface/seabottom position reference
systems includes:

-   Tracking of towfish, maximum distance 2 km.

-   Tracking of seismic streamers, streamer separation (3-D seismic),
    maximum distance 3 km.

-   Tracking of underwater vehicles from thruster operated vessels up
    to 1 km in shallow (10 m) water.

TABLE 5. Characteristics of SSBL position reference systems.

CHARACTERISTICS	SIMRAD HPR SYSTEMS HPR 309	THOMSON SINTRA SYSTEMS TSM 7321 (1)
OPERATING FREQUENCY	21-32 kHz	18-25 kHz
TRANSDUCER BEAMWIDTH	MECH./ELECT. 30°	FIXED 60°
MAXIMUM RANGE	2500 m	750 m
ACCURACY OF SLANT RANGE	< 1 m	< 1 %
Nb OF TRANSPONDERS	16	2

TABLE 6. Acoustic transmission system.

CHARACTERISTICS	UNDERWATER TELEMETRYSYSTER				
	SIMRAD			THOMSON SINTRA	
OPERATION	ACOUSTICAL CONTROL	SUBSEA TEMPLATE OPERATION	CONTROL LINK FOR ROV	TV UNDERWATER TRANSMISSION TSM 5555	HIGH SECURITY TRANSMISSION TSM 5556
FREQUENCY	20-33	20-24	65-70	60	18
BEAM SHAPE	100° CONE	13° CONE	180° OMNI	60° CONE	60° CONE
MAXIMUM DISTANCE	1000 m	8000 m	300 m	1200 m	1200 m
TRANSMISSION RATE	2 BAUD	2 BAUD	50 BAUD	20000 BAUD	70 BAUD

As an example, characteristics of two SSBL-systems are given in Table 5. Special-purpose configurations of navigation/positioning instrumentation include:

- Deep-sea positioning (6000 m).

- Deep-sea local navigation (sub-meter range resolution).

- Portable instrumentation for ROV-handling.

In Figure 8 an application for open-sea tanker-loading is shown, where crude oil is loaded by use of flexible pipes, a sub-surface buoy, 50 m below surface, ensuring necessary tension on the vertical riser. The positioning of the tanker, relative to the buoy and its gravity base, is monitored acoustically, as well as status information of valve operation and buoyancy of the riser structure.
In Figure 9 instrumentation for deep-sea mining is shown. Different acoustical techniques may be applied for dynamic-positioning of surface vessel (low-frequency long - baseline technique) and riser-angle monitoring by underwater telemetry.
Positioning instrumentation are implemented in larger ocean systems, like monitoring systems for acoustical tomography.

5.2. Subsea Monitoring and Control

Table 6 gives different telemetry instrumentation applied in the offshore market. The SIMRAD systems include:

- High-reliable transmission systems for BOP (Blow-Out Prevention), control, status signal handling.

- Long-distance (fixed point-to-point) template monitoring and control.

- Compact and portable instrumentation for communication with (autonomous) vehicles.

The THOMSON Systems include a 20,000 Baud acoustic link for TV UNDERWATER TRANSMISSION.
The combination of control/telemetry links (ACL/ATL) with local horizontal/vertical acoustic detection systems, as illustrated in Figure 10, are now in development.
The systems described above will qualify as general-purpose instrumentation for ocean monitoring systems.

6.   CONCLUSIONS

Commonality between ocean systems and systems applied in naval and hydrocarbon markets can be demonstrated by use of complementary acoustical measurement techniques and will give a framework for developing commonality in design/technology.

The authors have focussed on the following categories:

- Sidescan sonars,

- Multibeam echo sounders,

- High resolution sonars for underwater vehicles,

- Systems for underwater positioning/navigation,

- Subsea control and monitoring,

describing areas where technology transfer does exist and should be promoted.

High performance/reliability are ultimate technology goals for naval/offshore acoustical instrumentation. Technology transfers across borders and into ocean system applications will be highly beneficial and cost-effective for bringing the next generation of technology in EEZ-related applications.

PART IV

Future Technology Requirements

THE IMPORTANCE OF OCEAN DEVELOPMENT TO NEWLY INDUSTRIALISED COUNTRIES
AND LESS DEVELOPED COUNTRIES

K.T. LI
Senior Advisor to the President,
R.O.C.

1. INTRODUCTION

It gives me great pleasure to address this inaugural meeting of
the International Ocean Technology Congress. I feel it is most
appropriate that you have chosen EEZ development as the main theme of
this conference. As one senior government official of the United
States has stated, and I wholeheartedly agree with her, "the Exclusive
Economic Zone is our backyard equivalent to outer space - a new
frontier on Earth. It is a challenge for our technology, for science,
and for the mind".

The potential wealth of the sea was brought to international
attention by the issuance of the EEZ proclamation, which has been
signed by over one hundred nations since 1982. In assessing the
resources within the EEZ, a country must take a phased approach
involving co-ordination among government agencies, academic
institutions and the private sector. For newly industrialized
countries such as mine, international collaboration through the
transfer of technology and exchange of information with developed
countries is essential. For less developed countries, financial
support may also be required.

As you are all aware, the relationship between man and sea dates
from the beginning of human existence on this planet. Mankind has been
securing food from the sea, and using the sea as a means of livelihood
and transportation since the dawn of civilization. Although
traditional fishing methods are still used in some less-developed
coastal states, with the tools of modern science today's fishermen are
able to meet the food needs of entire nations. The industrialized
countries are beginning to exploit the mineral resources of the sea,
including hydrocarbon resources, and to utilize sources of energy which
abound in the world's oceans. In all these areas, the gap between
developing and developed countries is widening. Even the newly
industrialized countries do not have sufficient technological
capability to tap the resources of their own EEZs or to assist other
countries in ocean development.

I use the term "ocean development" advisedly, because the
traditional method of fishing is rapidly giving way to more modern

D. A. Ardus and M. A. Champ (eds.), Ocean Resources, Vol. II, 217–223.

techniques, and new and innovative technology is creating a revolution
in marine transportation and the utilization of ocean space. Indeed,
man's expectations of the ocean are continuously rising in response to
his pioneering spirit and to the need for resources and new space due
to increasing population. In fact, it is no exaggeration to say that
ocean development could be the key industry of the 21st century.

For the reasons I have mentioned, the holding of this
International Ocean Technology Congress occurs at a particularly
appropriate time, and the EEZ, as our backyard, should receive our
priority consideration.

To fully utilize the EEZ, we need more reliable estimates of
resource potential and greater knowledge of oceanic processes
pertaining to the continental shelf and slope. While basic
oceanographic research remains important, advances in marine technology
and marine engineering should also be closely watched.

Satellite remote sensing has played a significant role in the
study of global marine processes and, we hope, some day space platforms
will be able to perform this function. Manned and unmanned
submersibles and remotely operated vehicles are undoubtedly needed to
provide a close view of the ocean bottom. Initial surveys of large
areas require newly developed mapping systems, which are presently
beyond the reach of the newly industrialized countries and less
developed countries. Better fish-finding capabilities and improved
navigational positioning systems will also be required to enable users
to conduct surveys more efficiently and accurately.

In the exploitation and development of the EEZ, the issue of
environmental impact needs to be addressed. The question of
environment vs. development received the early attention of developed
nations. It is now becoming a serious problem for newly industrialized
countries and will, eventually, be of major concern to less developed
countries also. These countries do not have sufficient background data
and baseline information to make wise and well-reasoned decisions
regarding the development of EEZ resources. In a number of cases, the
natural baseline has already been altered by past human activities.
This baseline information needs to be recorded and evaluated for
licensing and long-range planning, as well as for possible litigation
purposes in the future.

The issues I have just touched on are, of course, common
knowledge, and I don't think it would be difficult to convince even a
layman of their importance. However, recognizing the importance of
ocean development is one thing: actually implementing a development
program is far more difficult - especially for the developing or newly
industrialized countries. By and large, these countries not only lack
fundamental data about their EEZs, but are also short of the necessary
expertise and technology. In addition, the statutory and regulatory
frameworks needed for the orderly development of the EEZ are largely
non-existent in these countries.

In view of the circumstances, I have here described the newly
industrialized countries and less developed countries are looking to
this congress for new information and guidance. For the same reason, I
have brought with me as participants a number of ocean scientists,

ocean engineers and others from my country who are concerned about the
ocean environment.

Ocean technologies encompass a wide field. Permit me to mention a
few which are particularly relevant to the development of the Exclusive
Economic Zones of newly industrialized countries and less developed
countries in the Asian-Pacific region. These technologies can be
divided into four broad categories, namely, those pertaining to the
exploitation of ocean resources, coastal ocean-space utilization, ocean
engineering and the security of the ocean environment.

2.   OCEAN RESOURCES

In the exploitation of ocean resources, the first priority that comes
to mind is the development of **fisheries.** The proclamation of 200-mile
Exclusive Economic Zones has resulted in a 35% reduction in the area of
ocean open to international fishing operations. Offshore fisheries may
be an answer to this restriction, but there is a natural limit to the
production capacity of offshore fisheries. In order to more
effectively utilize the now more restricted sea area available for
fishing, the restructuring of fisheries and/or better international co-
operation through official agreements will be required.

In the context of the EEZ, the most serious problem newly
industrialized countries and less developed countries are facing is the
lack of managerial technology needed to improve the fish catch from the
most productive shelf waters. Better management of fisheries resources
is often taken to mean setting quotas and scaling back fishing efforts
based on, sometimes, rather poor and inadequate information about
fisheries stocks. This interpretation is incorrect and misleading.
Improvement in fisheries productivity can be achieved through better
understanding of the dynamics of fisheries stocks and the reactions of
fish to changes in their biological and physical environment. This
better understanding will make possible the optimal scheduling and
organization of fishing operations and the development of methods to
enhance the quality of the natural stock.

Countries in the Western Pacific are paying considerable attention
to **offshore hydrocarbon exploitation.** Southeast Asia has substantial
proven offshore gas reserves. Offshore petroleum exploration in that
area has met with some success, and there is hope of further
discoveries. The exploitation of offshore hydrocarbon deposits,
however, involves prospecting, drilling and production techniques as
well as scientific knowledge and engineering skills, which are still
beyond the reach of many newly industrialized countries and less
developed countries and which must therefore be provided by the
industrialized countries.

Ocean thermal energy conversion, or OTEC, is generally considered
to be the most promising technique for utilizing ocean energy
resources. It is promising because it can provide baseload electricity
at no fuel cost, it has only benign effects on the environment, and it
generates a number of valuable co-products including fresh water. OTEC
can be particularly important to tropical Pacific island nations whose
fresh water and electricity are in short supply and whose economies

depend on marine resources. I am therefore pleased to note that
several papers presented at this meeting will deal with OTEC.
    We in Taiwan have been giving considerable attention to OTEC.
With the ocean floor sloping steeply away from our east coast, the
temperature differential between surface and deeper water is greater
than that normally required for OTEC and an OTEC plant would easily be
constructed on shore. We are prepared to make our OTEC project an
international undertaking, and would welcome the participation of
foreign experts. If, through such a joint effort, we succeed in
establishing an OTEC plant of minimum commercial size, the experience
and knowledge gained could be passed on to island states located in
subtropical and tropical areas. In short, such a project would provide
an excellent opportunity for collaboration among developed countries,
newly industrialized countries, and developing countries.
    Having briefly discussed marine resources, with particular
reference to the EEZs of newly industrialized countries and less
developed countries, I would now like to say a few words about **marine
biotechnology**. Newly industrialized countries and less developed
countries are paying considerable attention to, and have already
started work on, land-based biotechnology research and development.
However, the diversity and abundance of marine organisms are far
greater than those on land, and newly industrialized countries and less
developed countries have not yet reached the stage where they can mount
a serious effort in marine biotechnology research and development. If
a system could be established for the industrial production of the many
types of organisms which exist in the sea, a new frontier in
biotechnology could be opened. This development would contribute
significantly to the upgrading of the industrial structure.

3.   COASTAL OCEAN-SPACE UTILIZATION

Coastal ocean space includes the region extending from the hightide
level of the shore out to and including the continental shelf located
within the 200-mile Exclusive Economic Zone. Increased population
growth and rising demand for coastal ocean products and services have
created a need for the more rational use of ocean space. This involves
marine transportation and port and harbour facilities, submarine
pipelines and tunnels, land reclamation, and industrial space.
    Shipping congestion is a problem facing many coastal states,
including newly industrialized countries and less developed countries.
Harbours have undergone dramatic change, owing to the increase in size
and specialization of freight shipments and the emergence of large
tankers, large ore vessels, and container ships. Planning for the
redevelopment of harbours and their hinterlands, water front
development, and other related activities should be given high
priority.
    Submarine pipelines are important for countries producing offshore
hydrocarbons in the Asian-Pacific region, and for the transportation of
oil and natural gas from production sites to consumption sites.
Submarine pipelines are also important for the intake and the discharge
of cooling water for power plants, and for the discharge of waste water
from households and industrial plants.

The reclamation of offshore space, including tideland and
nourished-beach reclamation, has been carried out in many countries,
including my own. Such an undertaking is important, particularly for
small island countries with high population densities. Such reclaimed
space serves as farmland, fish ponds, salt fields and, of course,
industrial sites. We will also have to take into consideration
regional development, including improvement of the environment for
dwelling, transportation, recreation and other activities. Of course,
utilization of ocean space can be achieved not only be filling in
coastal land zones but also by employing fixed or floating marine
structures, a subject we will leave the developed countries to
consider.

4.   OCEAN ENGINEERING

In the utilization of the ocean, we cannot ignore ocean engineering,
which is the application of oceanographic knowledge and engineering
skills to the development of marine resources and ocean-space
utilization for the benefit of mankind. It is the application of many
existing engineering and scientific laws to the ocean environment,
resulting in the improved design, analysis, development, construction,
deployment, operation and maintenance of ocean structures and systems.
        We generally divide ocean engineering into three main categories:
coastal, offshore, and deep ocean. As far as newly industrialized
countries, less developed countries and the purposes of this conference
are concerned, the first two categories deserve our particular
attention.
        Coastal engineering addresses all ocean-related problems in the
area extending from the surf zone to the coastline. Offshore
engineering operations are conducted in the area extending from the
surf zone to the continental slope, and are mainly concerned with the
exploration and development of hydrocarbons, energy, and mineral
resources.
        When we talk about fisheries, we think of marine ranching and such
new fields as "aquacultural engineering". The commercialization of
aquaculture has made it increasingly clear that engineering technology,
in its broad sense, is the required catalyst that ensures efficiency in
aquacultural production.
        When we consider offshore hydrocarbon resources, drilling rigs,
production, and transportation come to mind.
        In the utilization of ocean energy, we take into consideration not
only evaporators, turbines and generators, but also undersea pipelines,
which are, indeed, the most expensive part of OTEC development.
        In the utilization of coastal ocean space, harbour construction
and improvement deserve our attention. Harbour structure is a special
field of its own consisting of wave protection, mooring, and loading
and administrative structures. There are also coastal structures for
protecting shorelines, such as revetments and breakwaters.
        It must be realised that all these activities require specialized
equipment and construction tools for use at sea, and trained
technicians and engineers to obtain reliable data and do an efficient
job. Equipment and tools must also be maintained, repaired, upgraded

and replaced. Because of its importance, I consider ocean engineering one of the four main categories of ocean technology. As far as I know, basic oceanographic research has been given attention by newly industrialized countries and even less developed countries, but the engineering side seems to have been less emphasized. I feel that this situation should be corrected.

5. OCEAN ENVIRONMENT

At the beginning of this speech, I mentioned that environmental impacts should always be kept in mind in the exploitation of ocean resources and ocean-space utilization, particularly in EEZ development.

Having participated in the development of my country for over forty years, I in no way oppose the development of the EEZ. On the contrary, I strongly encourage it. However, I am deeply concerned about marine pollution.

The continental shelves provide the ultimate sink for the byproducts of human activities. The shelves receive wastes from cities, farms, and industries, and even seemingly innocent recreational facilities. As a result, coastal waters along the world's shorelines have shown signs of stress from the billions of tons of contaminants added each year. In order to sustain the growth of the world economy and the burgeoning demand for food, fuel, transportation and ocean space, the use of coastal areas for settlement, industry, energy and communication facilities, aquaculture and recreation will certainly accelerate, as will the upstream manipulation of estuarine river systems and the unwise and unregulated use of groundwater. These activities have frequently destroyed estuarine and coastal habitats as irrevocably as direct construction activities. Man-made structures also induce changes in current flow patterns, causing coastal corrosion and resulting in changes in natural marine ecosystems.

Certain coastal and offshore waters are especially vulnerable to ecologically insensitive onshore development, to over-fishing, and to pollution. These trends are of special concern in coastal areas where pollution by domestic sewage, industrial waste, pesticides and fertilizer run-off, and ocean outfalls and dumping may threaten not only the marine biota but also human health. High concentrations of pollutants have been found in localized regions. Whether accelerating pressure for development would damage more food chains, disrupt more planetary support systems, and enhance coastal erosion, needs to be assessed. I would like to call your attention to this important "Land-Sea interface" problem.

The need to utilize ocean resources and ocean space, and the concern about harming the ocean environment call for a unified Land-Sea interface management approach. It is also important to bear in mind that sound management of ocean resources requires proper management of land-based activities as well. Inland areas affect the oceans via rivers, and human activities on coastal lands affect the adjacent waters directly. Land subsidence, seawater intrusion, flooding, and enhanced coastal erosion have been reported in areas with over-use of groundwater. Shore protection in these areas has become important.

All these factors point to the need for treating and managing the
oceans, the shore, and the land as one integrated unit.

I have witnessed and participated in the evolution of an economy
from the developing to the newly industrialized stage. My remarks are
therefore understandably confined to ocean development in newly
industrialized countries and less developed countries. The topics
which I have put forward are meant to be illustrative rather than
exhaustive. I have not touched on more sophisticated technologies
which are beyond the capability of less developed countries and newly
industrialized countries at present and whose adoption is considered
less urgent.

International conferences have provided excellent forums for
countries to exchange information on their activities, their
achievements, and their policies. They have also enabled experts and
decision makers to get acquainted with each other and to discuss
matters of common interest.

It is my hope that your Congress will achieve more than this.
This Congress should be action-oriented. It should help newly
industrialized countries and less developed countries to formulate
ocean policy and identify areas for priority consideration, and should
otherwise assist them as far as possible. In this connection,
international co-operation among developed countries, newly
industrialized countries, and less developed countries is very
important, regardless of ideological or political differences.

I have been told that Canada, through its International
Development Agency, has assisted Indonesia in the formulation of an
action plan for the sustainable development of Indonesia's marine and
coastal resources. As a newly industrialized country, the ROC has
established an overseas economic co-operation development fund. This
fund will ultimately have at its disposal an amount equivalent to
around one billion US dollars within a specified period in co-operating
with international development financing agencies, which will be used
to help less developed countries develop their natural resources,
including marine resources. I am therefore pleased to note that,
towards the end of this Conference, an International Collaboration
Workshop will be held to discuss multi-nation co-operation on EEZ
projects.

FUTURE OCEAN ENGINEERING, SUBSEA WORK SYSTEMS AND TECHNOLOGY
REQUIREMENTS AS RELATED TO OCEAN RESOURCE DEVELOPMENT

JOHN P. CRAVEN *
JAMES G. WENZEL **
MICHAEL A. CHAMP ***

1. INTRODUCTION

The declaration of Exclusive Economic Zones (EEZ) increases greatly the
area over which coastal nations have exclusive control of the resources
of the earth. Unlike the land jurisdiction, the title to the resources
of the Exclusive Economic Zone and the terms and conditions under which
title passes to private ownership, vest in the people of Nations
through Federal and State sovereignty. The resources of this vast
domain and the manner in which they will be exploited is as yet
unknown.
    Many engineers will believe that the problems of resource
development in the ocean will be straightforward on the presumption
that the sophisticated technologies developed for the land, the
atmosphere and space will find immediate and direct application in the
ocean. But those engineers having extensive empirical experience with
the ocean and the deep ocean will recognise that the physical nature of
the ocean environment is such that engineering solutions, developed for
use in other environments, are usually inappropriate, are often
counter-productive, and on far too many occasions catastrophic.
    The development of low-cost, safe and effective solutions to the
fundamental problems of the generation from the sea of food, fibre,
energy, transportation, the occupation of ocean space, recreation and
the development of aesthetic and cultural enjoyment opportunities,
climate control and weather modification, enhancement and protection of
the environment, will require a new and widespread understanding of
engineering in the ocean as an **ocean system** integrating the physical,
chemical, geological, and biological properties of the sea.
    The next era of ocean engineering will probably be referred to as
ocean systems engineering. Acceptance of this thesis requires a
historical perspective. Prior to World War II the settled and accepted
use of the sea was for the international transport of peoples and
materials, the extension of military and political power and influence,

*   The University of Hawaii
**   Marine Development Associates Inc.
*** Environmental Systems Development Inc.

D. A. Ardus and M. A. Champ (eds.), Ocean Resources, Vol. II, 225–240.
© 1990 Kluwer Academic Publishers. Printed in the Netherlands.

and for the limited harvest of fish from a relatively inexhaustible supply. The War itself, a struggle between Democracy and Fascism was won by the development of three technologies, which controlling the skies, controlled the land beneath the skies; the submarine which, roaming unmolested beneath the seas, controlled the logistical use of the sea and the nuclear weapon which, visiting a unilateral devastation destroyed the will to resist. Other technologies experienced major advances. The use of the electron and electromagnetic radiation in radar and electronic systems played a major role in search and surveillance and in command, control and communication. In the post-war period the vast majority of these technologies found their application in land, air and space projects with but a few notable oceanic exceptions. Of these, most notable, was the marriage between nuclear power and the submarine which opened the undersea and under-ice to continuous occupation for protracted periods of time. Less notable but perhaps of greater significance in the development of the ocean was the development of the offshore platform for the mounting of radar picket installations for coastal defence. These were the forerunners of the offshore oil platforms which were soon to appear in exploitation of the oil and gas on the Continental Shelf.

The quantity of offshore oil and the attainment of a technology which, albeit unsophisticated, could recover it produced a dramatic change in the uses of the sea. Now the continental shelves beneath the ocean became a major source of crude and the oceans became the major avenue for the transportation and distribution of crude oil and petroleum products. The "strategic lanes of communication" or "SLOCS" so developed are clearly identifiable and their protection has become a primary foreign policy and defence policy goal of each administration. In a larger sense, todays use of the ocean is in the generation and distribution of energy and energy products, commercial shipping, in the production and harvest of marine protein and in the management of the disposal of selected societal wastes and it is to these more systematic systems developments that we must look in determining the most appropriate directions for the development of new ocean technologies.

In the United States, the development of ocean resources has been significantly hindered by a policy that has developed in the early 70's at the same time outer space exploration programs were being developed. The policy developed from the key Federal assumption that private industry would fund the technology development required for commercial uses of the ocean. However, this was not true for the exploration and development of outer space, which was initially extensively funded by public funds, and currently has a federal budget of over $20 billion. It is unfortunate, for this decision was based in some part on the ocean not being as attractive as outer space or mankind's travel in outer space, because the oceans hold an underwater world of resources far greater than imaginable. In 15 years of exploration of outer space, not a single resource has been brought back.

Also the research support in ocean engineering/technology was limited or non-existent by virtue of University refusal to conduct classified research and the paucity of ocean technology programs associated with other Federal Agency missions. Another missing link in the process (which was so effective during and immediately after World

War II) appears to be the failure of American Universities to educate a
substantial number of ocean engineers and engineers oriented to the
ocean.

Unfortunately, these examples demonstrate how a simple policy
decision such as non federal participation in the exploration and
development of the oceans resources can significantly effect national
trade balances, such as the $24 billion trade deficit in minerals in
1984 (the last year data is available). By the year 2000 the U.S.
trade deficit in minerals has been estimated to exceed $50 billion.

The impact of this policy was also demonstrated dramatically again
in the eighties, when heavy industry investment in the U.S., in ocean
engineering came essentially to a halt. Due to the low price of crude
oil, the offshore oil industry, which has carried the load of ocean
technology development since the 50's, was not able to afford to
continue such efforts. With the associated (and temporary) low-cost
energy, and the Department of Energy (DOE) major cutback in alternate
energy research, U.S. industry also has had to walk away from the
promising development potential of OTEC (Offshore Thermal Energy
Conversion). The Law of the Sea and the strategic metals market
collapse have terminated the aggressive deep ocean mining technology
developments of the seventies within industry.

The significance of ocean development to nations of the world was
recognized in the late sixties with the initiation of the United
Nations Third Conference on the Law of the Sea. This Conference
produced a text for the Law of the Sea which, in large measure, and
with particularlized exceptions, has been identified as a correct
codification of the customary Law of the Sea.

The value of ocean space and resources has been clearly recognised
by foreign nations such as West Germany, Japan, the United Kingdom,
France and Norway where fundamental ocean engineering research within
industry is continuing under Government support. Similar rethinking is
important to all nations if they wish to remain internationally
competitive in strategic materials, energy, and export-import
economics.

2.   HISTORICAL OCEAN ENGINEERING AND TECHNOLOGIES

For this Section, the following approach has been selected to discuss
ocean engineering and technology requirements for: surveys and
sampling, development of ocean space and resources, the oceanic
functions (transportation, navigation, communication, and work
[underwater machines and robotics]) from both a historical and future
perspective.

2.1. The Seabed-Survey

Seabed explorers have used a diverse array of tools in recent years.
For shallow water sensors deployed in orbit around the earth, in
conventional aircraft, and from the sea surface have had surprising
success. However, because of the opaqueness of the water medium,
surveys in deep water have had to employ acoustic, magnetic, gravity,
seismic and electromagnetic systems. To resolve many details and for

the highest resolution possible, submersibles, remotely operated
vehicles and towed systems, operating similar systems plus cameras and
high-frequency, low-range acoustic devices are required. It is not
possible here to describe adequately all of these tools, but only to
indicate general capabilities and to describe some limitations which
can be overcome by engineering development efforts. Reconnaissance
tools for the seabed and, in particular, acoustic systems are
considered.

Three systems illustrate the state-or-the-art in surface-deployed
deep seabed mapping: SEABEAM, SEAMARC II, and GLORIA. The following
brief introductions to each of these are presented to show the
potential these systems have to provide accurate maps of seafloor
bathymetry.

SEABEAM is a hull-mounted array of narrow-beam, low frequency
transducers, adapted from the U.S. Navy SASS system. SEABEAM is
currently in operation on a number of American and foreign
oceanographic research ships. It has already been used by the National
Oceanic and Atmospheric Administration (NOAA) in surveys of the U.S.
EEZ. The system is capable of generating bathymetric maps in real
time, covering a swath under the survey line about three quarters of
the local water depth. Ten metre contour intervals are produced, and
seabed features smaller than a few tens of metres can be distinguished,
if they have a similar vertical extent. The most severe limitation
currently with SEABEAM is related to navigation. As discussed,
conventional satellite navigation, presently the only means in many
areas for obtaining precise position data, can only provide occasional
fixes; for the periods between satellite passes, a dead-reckoning
extrapolation must suffice. Therefore, the ship's actual course and
speed are only known after the fact, and accurate maps must be produced
by post-processing of the SEABEAM data with the interpolated position
information. The uncertainties in this interpolation become very
evident when the investigator attempts to resolve the discrepancies
between adjacent swath maps. When continuous high-resolution
navigation methods are available, such as LORAN C or GPS, these
problems are essentially solved. Thus, with the full deployment of the
GPS satellites in the near future, the true power of SEABEAM will
finally be usable on a routine basis.

GLORIA is a low-frequency sidescan sonar system which is towed
near the sea surface. It produces acoustic images of the seafloor in a
swath similar in size to that of SEABEAM. GLORIA also suffers from the
same navigation-induced limitations as SEABEAM and, in addition, is
less available for general usage, given its unique nature and the
extensive post-processing that is required to produce usable maps.

SEAMARC II is also a one-of-a-kind side-scan system, developed,
operated and owned by the Hawaii Institute of Geophysics. SEAMARC II
is a higher frequency system, which produces more definitive
bathymetric maps of the seafloor, albeit over a narrower swath width.

2.2. The Seabed-Sampling Tools

Three types of samplers exist to retrieve samples of the seabed:
dredges, grabs and corers. Many varieties of each type have been used.

(1)   Dredges -
     Dredges are the simplest and yet most poorly understood of all
sampling equipment. Samples from a dredge are almost entirely
qualitative. Without complex modifications, the investigator has
little insight as to how much of the seafloor is represented by the
sample, or if it is representative at all. This results because
dredges preferentially collect samples which protrude above the ocean
terrain and skip over smooth flat regions. The development of dredges
which can overcome this problem and still retain the excellent
simplicity of the tools, would be a substantial boon to the seabed
explorer.

(2)   Grab Samplers -
     Next to the dredge, the grab sampler is the simplest sampling tool
available. The advantage of grabs is that they give a much more
representative idea of the seafloor than do dredges. The disadvantage
is that generally the area being sampled and the sample recovered are
much smaller. This disadvantage has been overcome in recent years, to
some extent by the development of free-fall samplers. These permit the
collection of many samples in a time period previously required for a
single sample recovery. The key area where development is needed for
grab samplers is in the capacity to retrieve samples from hard
substrates. Most conventional and free-fall devices are very
inefficient or completely ineffective at bringing back samples from
hard outcrops. Since many mineral resources in the ocean are expected
to be found on, or in hard substrates, this is a serious disadvantage
for seabed resource assessment.

(3)   Coring -
     To investigate the seabed below the immediate surface, coring is
essential. Gravity corers, piston corers, box corers and other
varieties currently do an adequate job of retrieving shallow subsurface
samples from soft substrates. The problem again is with the hard
substrates. Depth of penetration is also a serious limitation with
existing corers. Presently the only means of getting samples from more
than the top few metres is to wait for the Ocean Drilling Program to
select the candidate site for deep sea drilling operations. Given the
many competitive demands on the ODP, this may be a very long wait.
Wire-deployed or submersible-deployed corers, which can sample to
depths of a few tens of metres in hard substrates, would obviate the
need for deep sea drilling in many cases and would add significantly to
the capabilities of seabed explorers.

     Dredges, samplers and corers in use today provide at best crude
and qualitative information about the seabed. Developed on a
scientific project by project basis, they demonstrate that the
profession is literally at ground zero in the technological capability
to explore and exploit the seabed.

3.   FUTURE SUBSEA WORK SYSTEM REQUIREMENTS

The best way to sample an outcrop is clearly on-site. More insight has been gained about seabed geology, in recent years, through submersible operations, than through any other technique. For the deep seabed in particular, submersible sampling is unparalleled for control and ease of interpretation. The primary problem in using submersibles for this purpose is power consumption. Existing submersibles can only collect easily accessible surface samples with relatively weak manipulator arms. Coring requires generally far more power than can be spared from the life support and propulsion systems.

3.1. Undersea Power and Propulsion

The primary limitation to the use of undersea systems as the chief mechanism for avoiding the free surface is simply the non-availability of atmospheric oxygen. Thus, conventional air breathing power systems are not available, nuclear power is very costly and hazardous and non-air breathing submersible power systems have only an embryonic development. A further apparent limitation is that underwater systems that carry their own oxidant and fuel will have to carry approximately eight times the amount of oxidizer than the amount of fuel.

Unsatisfactory solutions to this problem have been in the form of the snorkel for conducting the atmospheric air to the underwater power plant. This may indeed be a solution for underwater systems which operate in the vicinity of a fixed platform and can thereby replenish the oxidiser at frequent intervals. In point of fact, underwater vehicles are capable of carrying a very large, neutrally buoyant cargo with only a minor reduction in speed, as compared with a cargoless vehicle. (Power varies as the Velocity to the third power (V^3) whereas Power varies only with the first power of the cross sectional area and with a fractional power of the Length). A trade-off study conducted in connection with the design of the NR-1 (a small nuclear powered deep ocean research submersible), indicated that a conventionally-powered submersible having the same dimensions and weight as a hypothetical submersible similar to the NR-1, would have a mission endurance of twenty days. Given the current state-of-art-of-fuel cells (fuel cells were developed for operation on the U.S. Navy Deep Submergence Rescue Vehicle) a technology program to develop a commercial fuel cell submersible powering system would probably be successful. An alternate to the fuel cell which has been recurrently under investigation is the Sterling cycle engine. This engine which operates on heat requires this more simply derived product of the combustion of an oxidiser and a fuel.

More sophisticated solutions have been offered which include the simple and surprisingly effective technique of buoyancy propulsion (the fuel is low cost concrete and the weight of the propulsion plant is essentially zero) and the more complex and as yet unproven "artificial gill", which extracts oxygen from the seawater. The extracted oxygen, combined with, e.g. use of a methanol fuel cell, has the potential of essentially eliminating the power system constraint on endurance performance of small autonomous underwater vehicles.

Within this context the need is evident for a number of development paths for energy storage and conversion:

(a) Develop more effective and efficient power generation, distribution conditioning and energy storage for unique ocean engineering applications.

(b) Develop high rate primary and secondary lithium batteries.

(c) Increase power density of deep-ocean fuel cells by development of light weight, pressure compensated reactant storage and conversion subsystems.

(d) Power conditioning control, protection and regulation equipment for systems using high levels of electric power delivered by cable.

(e) Develop stored chemical energy propulsion systems including conversion of shaft power to electrical power.

(f) Develop buoyancy propulsion modules for sustained and burst power.

(g) Develop techniques for efficient in-situ low energy extraction of dissolved oxygen from seawater.

3.2. Materials and Structure

The implementation of the power and propulsion developments will require complementary developments in high strength materials and structures in order to take advantage of the energy potential for negative and positive buoyancy in waters of great depth. At present, hull and structure advances have been dictated by a very conservative approach on the part of the United States Navy, which has attempted to make progress within the constraints of the use of high strength steels and a commitment to the precisely configured ring stiffened cylindrical hull. The Soviet Navy has moved to titanium and small research submersibles have been effective with aluminium, titanium and acrylic hulls. The potential with such materials as composites, glass and ceramic is enormous as is the potential for unique and less costly pressure hull configurations. The construction of a low-cost, one atmosphere hull with a 0.5 weight to displacement would open up the entire ocean for utilization at costs which are almost invariant or even inversely related with depth. Specific development thrusts in this area are:

(a) The use of composite structures both organic and metallic or combinations thereof - Offers significant advantage in the ocean to reduce weight, increase strength and to counter corrosion. As has been the case with monolithic materials, these constructions will be utilized in a marine setting based upon success on land without the appropriate modifications for adaptation to the marine environment. Almost nothing is known about the effects of marine

corrosion and electro-chemical phenomena on the strength retention
of composites. Enough history of problems in wet environments has
accummulated to suggest major problems. Thus a fundamental
research program to examine generic fatigue and deterioration
problems in a seawater setting is required.

(b) More efficient pressure resistant structures - Weight displacement
ratios of 0.5 for all depths. Development of fibre composites,
titanium (with emphasis on welding) and ceramics. This requires
an understanding of failure mechanisms, design criteria, non-
destructive testing, and manufacturing techniques. As an adjunct,
transparent materials such as acrylic and glass need to be
examined for both pressure hull and optical transparency. Low
cost materials such as concrete should be improved for bottom-
fixed housings or hyperbaric test installations.

(c) Non-destructive testing - Develop capability to perform in-situ
non-destructive test and evaluation of underwater equipment,
structures and facilities.

(d) Advanced Underwater Materials Development - Concepts and practices
for "engineering" materials for deep water and for long exposures
will need to be developed, both for the commercial metals and the
advanced technology composites. Methods to predict materials
behaviour, failure conditions, deformations and stress levels, a
life cycle performance will need to be further developed.

(e) Corrosion and Biofouling - Deep water effects on corrosion have
not been thoroughly studied and engineering practice in this
design category is weak. Parametric studies and practices for
handling the effects of deep water currents, oxygen and other
dissolved gases, pH, and biological factors will need to be
conducted.

For the above reasons and for many other reasons it is desirable
to be able to build low cost (probably unmanned), deep ocean pressure
hulls. Small scale versions of such hulls already exist in the form of
the benthos spheres. Many other hull configurations such as the
polycylindrical pre-buckled hull give promise of dramatic reductions in
the cost and complexity of deep ocean hulls. Materials such as carbon
reinforced plastics, glass, ceramics, and composites thereof have a
much greater promise than any of the metallics as materials for deep
submergence.

3.3. Information Management and Communications

At the present time the land based information transfer and
communication systems place heavy reliance on satellite high frequency,
electromagnetic transfer of information from network and nodal
distributions of high-information rate and volume computer systems.
The revolution results from the low-cost high information capacity of
such units as the personalized computer and the relatively (relative to

AM and FM frequency systems) high bandwidth and information carrying
capacity of the satellite. Under-the-water electromagnetics are not
available except by cable. The substitution of acoustics reduces the
information channel capacity by several orders of magnitude.

There are many technologists who believe that, because of its
clean signal and high information channel capacity, the fibre-optic
cable network will soon make obsolete the satellite system for trans-
oceanic and global communication. Be that as it may, the ocean's
opaqueness to electromagnetic radiation forces the direction of
underwater communication and information handling. The clear links are
via surface and atmosphere communications to the first available and
most convenient fibre optic system, with fibre optics for long distance
cabled and tethered systems. Sophisticated, highest-frequency acoustic
links for moderate to long distance communications, where cables are
inconvenient or not possible and optical for extremely short underwater
links. The development of such a system should open up the undersea
for all high speed data processing and information systems that are now
so ubiquitous on the surface. Specific thrusts in this direction
include:

(a)   Development of fibre optic underwater cables, repeaters and
       underwater connectors and simple deployment techniques.

(b)   Development of acoustic cable-free data and control links.
       Standardise underwater acoustic telemetry practices with
       development of reliable acoustic data transfer systems.

(c)   Develop short-range optical signal transmission. Distances of
       from 20 to 100 metres are possible depending on water quality.
       Development of rapidly pulsed blue-green lasers and application of
       image intensifiers and optically gated systems to minimize back-
       scatter.

(d)   Develop display technology which maximises efficiency of
       information transfer between signal processing device and the
       human operator in the marine environment. Thus, non-optical
       displays should be explored, such as auditory patterns or even
       two-dimensional arrays of tactile excitation applied to the skin.

3.4. Information Management - Supercomputing Models and Prediction

The development of supercomputing capability, only recently available,
will have profound impacts upon ocean engineering and technology and
upon its productivity. Based on existing VLSI technology,
supercomputer speeds approaching 1,000 MIPS perform calculations that
model physical processes on scales heretofore impossible. Within a
decade or so, these speeds, with the necessary attendant memory
capacity requirements, are projected to reach levels of 100,000 MIPS or
more, a two order of magnitude increase. The models then tractable
will be more accurate, finer scaled and more synoptic. Microcomputers
will have a more profound impact on ocean-related engineering than the
supercomputer. PC level CAD/CAM systems available today are the future

and are extraordinarily powerful. Combined, these computer and data management systems will revolutionize the research and practice of all engineering and will profoundly influence the development of ocean-related engineering technologies. The power of these systems, and the problems made tractable by them, provide an exciting context for the development of the resources of the EEZ. Computer simulations and models have been developed to predict estuary flows and dynamics, storm surges, modest levels of nearshore ocean circulation, meso-scale weather and climate, regional sea conditions (i.e. sea state), dynamic response of vessels, platforms, structures and underwater systems, and distributions of nutrients and chemical species in the ocean. Virtually all these simulations and models are first order or gross scale. Many are not able to function in real-time, and most suffer from a lack of structure and detail that make them operationally effective and useful. The needs for research and the opportunity to apply new techniques are exciting and essential.

3.5. Information Management - Data Collection and Processing

The opportunities and needs that the EEZ poses for fundamental engineering research and new technology developments in the fields of instrumentation and measurement, data acquisition, storage and advanced information systems and data management are both exciting and demanding. Areal and time scale demands of the EEZ places special requirements on research in these engineering fields of study. Enormous amounts of data need to be structured for ease of acquisition and use, e.g. the oil industry has over 10 million track miles of subsea profiling and bathymetric data for the EEZ. Data bases of this magnitude will require new concepts for data management and storage, and new strategies for dissemination of such information to users.

3.6. Intelligent and Remote Underwater Work

The technology to decouple from the free surface and to provide power below the surface will make major advances in the feasibility and cost of many of the systems which have been cited. A concomitant technological development which is required is the adaptation of modern information processing equipment and the associated control, anthropomorphic control, programmed control and artificial intelligence control which is required for the submersible equipments to do sophisticated work beneath the sea. To the uninitiated this technology would be a simple adaptation from land and space systems. In point of fact the "added mass" aspects of the dense ocean medium result in control response characteristics which are long in comparison with human anticipation and analysis capabilities and satisfactory solutions will not be obtained without incorporation of the non-linear equations of hydrodynamic motion and the inclusion of sensory devices which measure the dynamic character of the local oceanic field. There is thus a substantial agenda in the development of "smart underwater systems".

    This underwater revolution begins with the explosive development of complex and often sophisticated microcomputer and micro-electronic systems, and is just beginning to impact ocean engineering and deep sea technology. Microcomputers, their attendant micro-electronics, and powerful soft/firm-ware provide the potential to address deep water and remote operations, heretofore either difficult or impossible. Acoustic tomography, sophisticated navigation systems, underwater imaging, extended range and resolution, remotely or autonomously operating underwater vehicle systems, and complex "smart" instrument systems are all examples of the impact this evolving technology is having on the ocean engineering. While these examples suggest important applications of "smart" and intelligent underwater systems, the full potential has not been exploited, primarily because applications to-date have been largely in pre-programmed systems, with virtually no capacity to adapt to changing conditions and environments. Knowledge-based systems (KBS) are conceived with the idea that they do adapt and accommodate such changing conditions and environments. KBS computer systems solve difficult problems, by using strategies and protocols which analogize, or even model, human expertise and insight, and which eliminate the need for massive search techniques. This development provides the framework for attacking real time systems. Real time KBS is projected as a keystone concept for ocean applications. The concepts and principles underlying KBS appear to provide an avenue of research for many ocean research opportunities and problems, with the potential of cost savings, savings in time, or making tractable the intractable.

A few potential examples suggest the range of research and technology development possibilities:

(a) Smart systems for managing remote oil and gas operations with complex equipment that is unmanned and operates unattended for long periods of time.

(b) KBS-based signal processing and pattern recognition optimise the use of massive amounts of data and minimises the need for extensive operator involvement.

(c) Smart ocean instrument and data gathering systems that operate unattended and adapt to changing environmental conditions.

(d) Autonomous free-swimming vehicles including work systems, detailed area survey vehicles, inspection of submersibles, and platforms for complex measurements (e.g. multistatic arrays to acoustically image the water column).

(e) Unmanned, autonomous surface vehicles and platforms.

(f) Complex experiment co-ordination and management.

(g) Identification of ocean processes and patterns.

Advances in computer hardware and software and the maturity of selected artificial intelligence and robotics concepts make possible the application of very intelligent autonomous work systems, platforms, vehicles and tools to the needs of offshore and ocean-based industries.

The real challenges lies in the ability of the research and technology
development community to "engineer" the principles of artificial
intelligence and robotics into real-world applications and
environments. The evolving field of knowledge engineering and the
capability of the essential computer systems make possible substantial
progress in intelligent and remote underwater ocean systems.

3.7. Robotics, Teleoperators and Subsea Work Systems

To develop and integrate the devices and techniques required to
supplant man and improve human performance in the ocean: (a) Controls
are needed for the development and integration of supervisory control
techniques, preprogrammed control and artificial intelligence,
(b) Sensors (visual, position, touch, force, proximity and temperature)
should be developed and adapted for use in future systems since visual
and position sensors are critical to most operations using either
manual or programmed control techniques, (d) Manipulators - three
classes to be developed: Lightweight (0 to 23 kg of lift capability)
manipulators of high dexterity for sophisticated precision tasks; Heavy
duty (23 to 91 kg) required for general work operations; Large
construction arms (91 kg plus) will be required for large seafloor
construction tasks.

3.8. Seafloor Construction and Installation

Finally, the completion of the preceding functional capability
developments will open the way for the construction of operational
systems facilities and sites on or near the seafloor. A whole range of
engineering capabilities generally covered under the rubric "Seafloor
Civil Engineering" will be required. There has already developed, both
in academic centres and in industry, an offshore engineering discipline
which has evolved an approach to addressing ocean and offshore
problems. This field of study will need to expand its frame of
reference, and address additional and new classes of research and
technology development challenges. Deeper waters, more complex and
severe environments, and remoteness from shore will drive much of the
new focus for research.
      Some examples suggest some of the research directions and trends:

(a)   Applied Fluid Mechanics and Hydrodynamics - Fundamental
      engineering research and the development of new technology
      concepts will evolve from basic fluid mechanics and hydrodynamics.
      The wide areal extent of the EEZ requires that new questions be
      addressed such as: (1) Non-linear wave mechanics and their
      interactions with structures and the seafloor, design
      methodologies and practices for both deep and shallow water waves
      and currents, engineering methods for predicting the influences of
      breaking waves on systems and structures, (2) Assessing the
      dynamic behaviour of compliantly-moored offshore systems,
      predicting flow-induced vibrations and dynamics, establishing
      operational models for predicting excitations, life cycling,
      fatigue, and dynamic responses, and (3) Developing methods for

modelling offshore wave fields both as a function of seasonality and areal extent.

(b) System Design for Increased Reliability - Remotely located systems (either deeply submerged or far offshore) require designs with higher reliability and greater structural integrity than current practice, which will require new improved understanding and engineering design practices for deepwater applications will need to address such topics as the prediction of joint and system stress concentrations, fatigue levels, and overall system static and dynamic behaviour under extreme loading conditions. Design strategies will need to be developed to predict the behaviour of operational systems, pipelines, and structures, including construction practices, welding technologies, and emplacement protocols, all with the over-riding demands imposed by deep and remote environments.

(c) Geotechnical Engineering - The deep waters and remote locations place special demands on the theory and practice of geotechnical engineering. Established practices in this field can not easily be extended into these new offshore environments. Methods and new design practices will need to be developed to assess and classify the seafloor foundations, and techniques will be required to predict the long-term stability of these special deep water environments.

(d) Synoptic Remote Sensing and Measurements Systems. Until the flight of satellites, there were no techniques for long-term, large scale, synoptic measurements of process in the atmosphere, oceans and solid earth. We are now on the verge of establishing a system of remote sensing instruments and earth-based calibration and validation programs that could provide such a data set. Remote sensing validation, and the related theoretical bases have expanded dramatically in the past few years, with wide bands of optical and electromagnetic radiation remote sensing capabilities. Research opportunities and needs exist for new satellite sensor technology, data processing, and information dissemination and management, i.e. needs for real-time information, fine scale details, and subsurface oceanic behaviour and structure. New technology concepts and strategies will need to be developed in all these areas.

(e) Tools - Light weight, safe, quiet, low cost improved power/weight ratio tools which are adaptable to diver or remote systems. Depth capabilities for pyrotechnic torches, shaped charges, chemical milling and explosive welding should be reliably increased.

(f) Develop accurate analysis, design and production techniques to obtain reliable undersea cable systems for communications, towed and tethered applications.

(g) Analysis/Design - Current analytical methods should be expanded to include three-dimensional techniques to gain a better understanding of stress levels and relative motions among cable structural elements in order to predict cable behaviour and lifetime under various loading conditions. A retirement criteria, validated by design, fabrication and extensive testing, should be established using the prediction model.

(h) Fibre optics - New material technologies that take full advantage of the separation now allowed between power, telemetry, and strength functions in the cable should be explored.

(i) Connectors, Penetrators and Terminations - A data base of proven designs should be developed to improve the reliability of electromechanical penetrators and terminations. These designs should include connectors for very high voltage (over 3000 volts), unusually high mechanical strength, or synthetic strength member terminations, wet mating by submersible swivels, sophisticated structures integrating sensors into cables directly, and other special technologies that are not available as standard commercial items.

(j) Repair - A splicing and repair capability including field repair, should be developed for complex electromechanical and electro-optical mechanical cables and terminators.

(k) System dynamic response prediction - Develop technique and analytic methodologies to specify and predict accurately system and environmental dynamics for ocean engineering applications.

(l) A dynamic analysis methodology that is sufficiently general, but includes appropriate hydrodynamical equations of motion and pressure distribution should be developed for a variety of applications. This capability should be extended to provide quasi-real time analysis of system dynamics. This will permit accurate design and planning to enlarge environmental operating windows and improved safety.

(m) The capability to predict spatial motions in real time must be developed and verified.

(n) The development of ocean current prediction and measurement techniques or models should be pursued, as these techniques or models must account for the complex interaction between the water mass, the system and the bottom topography.

(o) Load Handling - Develop techniques to permit the analysis, prediction and design of controlled handling, for any type of load, both on the surface and at any depth in the world ocean. An intensive effort should be made to understand and program the dynamics of load handling in a marine environment. This is particularly important since it will be necessary to predict some

forces and motions in a Lagrangian frame of reference and some in
an Eulerian frame of reference.

(p) Load Control - A concerted effort should be made to develop the
     theory, techniques and equipment necessary to permit any degree of
     desired load control. Of vital importance, is the differentiation
     and transfer between the three regimes; above, below, and at the
     air/water interface. Design goals should point to routine
     operations in sea state 5, moderate gale (30 knot wind speed) and
     4 knot current conditions. Subsurface conditions to be met are 4
     knot currents and internal wave heights of 1.2 to 2.4 metres.

(q) Moorings - Develop analytical and mechanical/design capabilities
     to improve the efficiency, operational life and load capacity of
     moorings and cable structures at all ocean depths. Expand
     analytical capabilities to predict holding capacity of anchors,
     dynamic response, strains on cables, etc.

(r) Mooring and Cable Structure Installation - A concerted effort
     should be made to develop and integrate the technologies which
     support reliable installation of mooring and cable structures at
     all ocean depths. The technologies centre on the equipment needed
     to transport and handle the hardware of the mooring cable system.

4.   SUMMARY

Fundamental engineering research, by its very nature, most often
focuses on detailed engineering concepts and theories, and functions
at levels of analyses that are deep within an overall engineering
system. Much of the research thoughts outlined above fall into this
framework. At another level, technology development must be set in the
context of overall system requirements and needs. Research programs
for the EEZ must always be driven by such, and new system concepts and
ideas should be fostered in parallel with fundamental engineering
research. It is out of such a mix that new technology emerges, and
effective and accepted new engineering practices are established. The
intellectual richness of the academic engineering research and
development community provides an exciting environment for developing
such new and creative ideas. The opportunity to develop and
responsibly exploit the EEZ, provides the reason.
     Many other technical developments could be cited (e.g. the
improvement of free-flooded machinery and components or underwater
construction, employing electrolytically deposited carbonates. However
the completion of the technological developments which breach the
barriers to "smart, decoupled, cheap and deep systems" will drastically
re-order the economic competitiveness and desirability of the cited
oceanic systems, as compared with their land based or land-oriented
alternatives.

5.   ACKNOWLEDGEMENTS

The material for this paper has been adapted from a conference and
workshop convened in 1986 by the University of Hawaii's College of
Engineering sponsored by the National Science Foundation entitled:
"Engineering Solutions for the Utilization of Exclusive Economic Zone
Resources" background papers; a reference publication, the Management
of Pacific Resources: Present Problems and Future Trends, by
J.P. Craven, Westview Press, Boulder, Colorado (1982); presented
papers, ideas presented and discussed by attendees during the
conference and/or workshop; and workshop summaries prepared by session
co-chairmen, the workshop Rapporteur, the conference co-chairmen, and
the staff of the College of Engineering of the University of Hawaii.
Sections of this manuscript were published in the Special Issue of Sea
Technology Magazine, June 1987: Vol. 28(6): 10-42.   The paper has also
been modified to reflect papers presented and discussions held at the
International Ocean Technology Congress (IOTC) Conference "EEZ
Resources: Technology assessment", in January 1989.